access to ɡeoɡraphy

D0588039

ECONOMIC ACTIVITY
and CHANGE

Paul Sheppard

DER
ATION
HETTE UK COMPANY

To all of the 'A' Level Geography students I taught
at Harlington School. To my parents, Peggy and Bert,
my beloved Hannah, Dusty now in the 'Spring-field',
and little Sam and Pedro

Acknowledgements

The publishers would like to thank the following individuals, institutions and companies for permission to reproduce copyright illustrations in this book:

www.baa.com/photolibrary, page 64. Peter Matthews/Folio Photography, page 70. *Advanced Geography: Concepts and Cases* by P. Guinness and G. Nagle, Hodder & Stoughton Educational, 2002, reprinted by permission of Hodder Arnold, page 44. Rio Tinto Mine, Southern Spain, Open-cast mine, Jonathon Kaplan/Still Pictures, page 96.

Every effort has been made to trace and acknowledge ownership of copyright. The publishers will be glad to make suitable arrangements with any copyright holders whom it has not been possible to contact.

Note about the Internet links in the book. The user should be aware that URLs or web addresses change regularly. Every effort has been made to ensure the accuracy of the URLs provided in this book on going to press. It is inevitable, however, that some will change. It is sometimes possible to find a relocated web page by just typing in the address of the home page for a website in the URL window of your browser.

Papers used in this book are natural, renewable and recyclable products. They are made from wood grown in sustainable forests. The logging and manufacturing processes conform to the environmental regulations of the country of origin.

Orders: please contact Bookpoint Ltd, 130 Milton Park, Abingdon, Oxon OX14 4SB. Telephone: (44) 01235 827720. Fax: (44) 01235 400454. Lines are open from 9.00 to 5.00, Monday to Saturday, with a 24 hour message answering service. You can also order through our website www.hodderheadline.co.uk.

British Library Cataloguing in Publication Data
A catalogue record for this title is available from the British Library

ISBN 978 0 340 81500 7

First Published 2004
Impression number 10 9 8 7 6 5 4
Year   2010 2009

Copyright © Paul Sheppard 2004

Cover photo: Ralph A. Clevenger/Corbis
Produced by Gray Publishing, Tunbridge Wells, Kent
Printed in Malta for Hodder Education,
An Hachette UK Company, 338 Euston Road, London NW1 3BH

Contents

1 An Introduction to Economic Activity

1 Introduction

The establishment of permanent settlements was the basis for economic activity, as people began to depend on each other for the provision of goods and services, and were not able to provide for themselves. Beginning 4000 years ago in the Middle East, the use of money instead of barter as the method of exchanging goods was a further stimulus to the development of trade. Economic activity can be seen as an integral part of any society from that time.

The economies of the nations of the world remained isolated for many millennia and it is only in the past 1000 years that the complex web of economic interaction in trade has developed within the Middle East, China and Europe, later spreading to incorporate the whole of the world.

The nature of this economic activity is dynamic, with the industrial revolution of the eighteenth and nineteenth centuries seeing the greatest change with regard to industrial process. This began in the UK, and later spread to Europe and North America, and brought about a whole new level of industrialisation. In the period since 1945, change has once again accelerated with the dawn of the computer age. Industry today remains an ever-evolving force that sees new countries gaining in industrial importance and dominance, sometimes at the expense of older more established industrial economies.

Similarly, the differing philosophies of capitalism, communism and social entrepreneurialism, which evolved in the nineteenth and twentieth centuries, have all had an impact on industrial activity, the individual and on societies as a whole. The beginning of the twenty-first century sees a greater emphasis on the capitalist philosophy, the impact of which is now experienced on a global scale.

2 Industrial Classification

Economic activity can be broken down into three sectors: primary, secondary and tertiary, with the latter being subdivided into quaternary and quinary, reflecting the technological revolutions that have taken place in the latter part of the twentieth century.

a) Primary industry

Primary activity can be classified as the first human contact with the environment, and is associated with farming, forestry, fishing and mining. These are extractive industries, and involve harvesting crops, rearing livestock, fishing from the seas or rivers, felling trees taken from the natural forests or plantations, or the extracting of minerals from the ground either by mining, quarrying or drilling.

These activities, especially farming and forestry, occupy the greatest proportion of the land's surface and employ differing proportions of the world's populations depending on the level of development within the country under consideration. LEDCs have a larger percentage of the working population employed in activities such as subsistence farming compared to MEDCs, due to the low level of capital and high level of manual labour required. However, alongside the small-scale farming operations sit plantations and cash-cropping in LEDCs where, for example, fresh produce, especially fruit and vegetables, is sold directly to the consumer and grain and cereals are sold to secondary industries for processing and refining before sale to the consumer. Even larger scale, highly mechanised agri-business operations exist in MEDCs.

The mining and extraction of the Earth's minerals have taken place for many thousands of years and the extent to which such minerals have been exploited has depended on technological advances, which have enabled resources, either exposed at the surface or concealed within the Earth, to be exploited. Early coal mining only occurred where the coal seams were exposed at the surface, and it was only with the development of deep mining techniques and the ability to locate such coal seams with the development of the geological sciences that concealed coal seams could be exploited. A resource only becomes a resource once it can be exploited or a use has been found for it. Similarly, the method of exploitation needs to be economically and technologically viable. For example, North Sea oil only became a resource once technology had devised methods for its extraction during the late 1960s and 1970s. This technology continues to develop, and oil companies are now capable of opening new offshore oil fields in deeper and more remote seas.

b) Secondary industry

The majority of primary products are then utilised by the secondary sector. It is at this stage that value is added to the product, either by refining it into a useable form or utilising it to make a final product. This sector of industry is traditionally the wealth-generating part of the economy as raw materials are manufactured into saleable products that have a higher value when processed. Raw materials such as iron ore are, by the process of smelting, converted into pig iron, and by the addition of oxygen into the smelter, steel. This steel is then itself a material used in the production of a range of products, often components, used in the construction of the final article. These are classed as capital goods as they are sold on to industry as opposed to consumer goods that are the final manufactured products, which are available for sale to the public.

Individual artisans used to dominate this sector of industry, but the technical or industrial revolution, which made the secondary sector of industry so dynamic since the eighteenth century, has made it dependent on machinery and increased the scale of production. This scenario is representative of MEDCs, but is not yet the case in many LEDCs where technology has still to make a large impact on industry.

The development of mass production by people such as Henry Ford in the 1920s had a further impact on secondary industry. Post-war developments dominated by greater efficiency, computer-controlled machines and increased output at a lower cost per unit all contributed to an increase in the size of this wealth-generating sector of the economy.

c) Tertiary industry

The tertiary sector of industry is concerned with the provision of a service. It includes the sale of the products of the primary and secondary sectors and the provision of services within societies. These include medical, teaching and social services, as well as retailing. It also includes financial and insurance services, which have gained in complexity as the technological revolution of the late twentieth century has evolved. The tertiary sector has become increasingly important as a source of employment as primary and secondary industries have become less important sources of employment. Greater efficiency and the use of machinery in these sectors in the MEDCs brought about this shift in employment opportunities. Demographic changes, for example an increasingly ageing population requiring more health care, and the rise in single occupancy households requiring more services, greater disposable income, more recreational time and the growth of local, national and international tourism, have all aided the growth of the tertiary sector of the economy. Staff to support and administer technological innovations, such as cash dispensers and telephone banking, have also helped growth in this sector.

In a three-sector model of employment, the proportion of the population employed in the primary, secondary and tertiary industries is an indication of the level of development within a nation. High levels of employment in the primary sector are indicative of an LEDC, e.g. Burkino Faso and Uzbekistan, while a small proportion in primary and higher levels in secondary and, particularly, tertiary industries are indicative of a MEDC, e.g. the UK and Australia. It is considered that as an economy develops and grows, the proportion of the working population employed in primary industry decreases and is transferred into employment in the secondary sector. This can be seen in the figures provided for China in Figure 1. As the economy further develops employment rises in the tertiary sector at the expense of the other two. Wealth generation is seen to be the factor lying behind this evolution. The rise in average income, passed on to the populace means a rise in disposable income to spend on services. This view was first postulated in the 1930s but can be criticised since such evolution was not seen as being 'for the masses' in the nineteenth century.

The data in Figure 1 depict trends in the percentage of population employed in each sector by countries regarded as being MEDCs or LEDCs. Countries such as the UK, Japan and Canada show low levels of employment in the primary sector, at 1.4, 1.4 and 2.3%, respectively, increased numbers employed in the secondary sector, at 24.9, 30.9 and 26.5%, respectively, but compensating for these low levels are the very high percentages of the population employed in the tertiary sector, at

Country	Primary (%)	Secondary (%)	Tertiary (%)
Australia	3	26	71
Belize	18	24	58
Brazil	8	36	56
Burkina Faso	35	17	48
Canada	2.3	26.5	71.2
China	15.2	51.2	33.6
Ecuador	11	33	56
Estonia	5.5	28.6	65.6
Japan	1.4	30.9	67.7
Malawi	37	16	47
Poland	3.8	35	61.2
Russia	5.8	34.6	59.6
Spain	4	31	65
UK	1.2	24.9	73.9
Uzbekistan	36	21	43

Figure 1 Primary, secondary and tertiary employment based on 2001–2002 data for 15 countries. *Source: CIA The World Factbook.*

73.9, 67.7 and 71.2%, respectively. On the other hand, LEDCs typically experience the reverse scenario, with high levels of the population being employed in the primary sector and much lower numbers employed in the tertiary sector, e.g. Burkino Faso and Uzbekistan. Low levels of the population employed in the secondary sector, or wealth-generating sector, compound the financial problems experienced by LEDCs. The figures also highlight China's emerging economy, with 51% of the population employed in the secondary sector. This information is displayed in Figure 2.

Clusters are produced in certain parts of the graph, which are indicative of the level of development of the countries the clusters contain. Geographers have always been interested in explaining such clusters and looking at how they have come about. This has led to many economic models, or simplifications of reality, being developed to explain what is seen in the real world.

One such economic model is that developed by W.W. Rostow, an American economist who devised a five-stage model to display the development and evolution of an economy over time. This is examined in the next section.

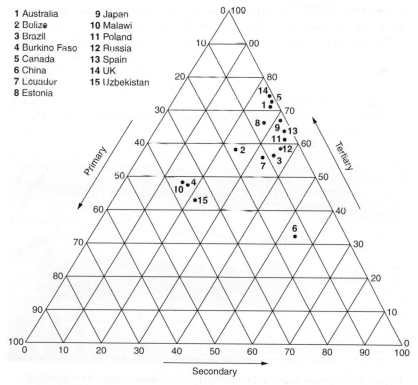

Figure 2 Economic structure of selected LEDC and MEDC nations.

3 Rostow's Model of Development

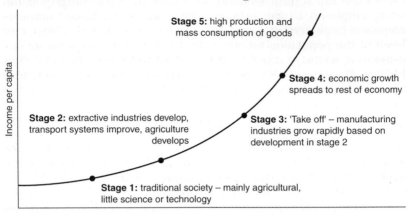

Figure 3 Rostow's stages of economic growth model.

a) Stage I: a traditional subsistence economy

The traditional society is one in which the concept or possibility of growth is minimal. There is a lack of applied science or technological innovation due to the lack of contact with the wider world and therefore the introduction of new ideas to the society is limited. There are few societies that fall into this category today, although tribes such as the Cuyabano in the Amazonian rainforest of Ecuador cling to their traditional way of life even though outside influence exists. This is

Figure 4 Cuyabano canoes. Traditional canoes hewn from tree trunks are being replaced with lighter aluminium canoes.

due to the army having outposts in the rainforest and the development of eco-tourism that brings in visitors to their reserve. It has meant that traditional materials such as wood have been supplemented by lightweight aluminium in the construction of boats and homes, and Western style T-shirts and clothing also compliment their usual attire. However, the Cuyabano are determined to maintain their traditional way of life and methods of hunting, which are still largely employed in their rainforest environment.

b) Stage 2: the pre-conditions for take-off

External links allow interaction with other people and the exchange of ideas occurs. The resources of an area begin to be exploited often by outside agencies in the form of colonial expansion or the involvement of a multinational company. The exploitation of the oil in the Ecuadorian rainforest or in the desert kingdoms of the Middle East are examples. The effect goes beyond the exploitation of the resource, as it is normally accompanied by urbanisation of the country, often with the dominance of a primate city and a developing infrastructure that enables both people and products to be transported more easily around the country and allows utilities such as electricity and water to be made available. This pattern of growth concentrates development in certain locations to the detriment of other areas of the country, enabling a developed core and an underdeveloped periphery, or edge, to develop.

c) Stage 3: the take-off to maturity

This is a period of sustained growth where the economy expands rapidly, especially in manufacturing. Growth becomes the norm at this stage, and to assist with manufacturing for the home market and export transport networks are constructed and improved. This not only includes roads, railways and airports but also harbours and pipelines to allow the import and export of produce. Expert assistance from other nations, often from governments or multinationals based in the MEDCs, arrives to oversee change. This may be accompanied by new political ideologies dedicated to change and democratise the host country.

Change is localised initially, but the ripple effect sees innovation spreading throughout the country. Ghana, in west Africa, had growth initially centred on the capital Accra. The completion in 1958 of the Volta Dam on the River Volta and the aluminium smelter at Akasombo saw industrialisation spread, but northern areas on the edge of the Sahel are still isolated and lacking the development seen elsewhere in the country.

At the same time, the agricultural sector is reformed bringing about an exodus of people, especially men, from the rural areas. They move to the towns and particularly the primate city in search of work, improvement and prosperity. Often the jobs available are low paid and the only accommodation is on the edge of the city or near the central business district (CBD) in overcrowded poor-quality housing. The nature of growth depends on the economic ideology existing at the time. The free-market principle of the Western world is not the only model of growth. The Eastern bloc, dominated by the former Soviet Union until 1989, saw growth guided by state planning and control. Poland and Estonia, former Soviet states, are examples of countries where the economy was developed as a result of central planning. The new town at Nova Huta, adjacent to the former Polish capital at Krakow, grew around the steelworks and housing in apartments was of a standard size with neighbourhood shopping facilities. In contrast, industrialisation in the newly industrialising countries (NICs) is planned to stimulate economic growth, but the planning is not as detailed as those countries where planning was on the basis of the socialist ideology that dominated eastern Europe.

d) Stage 4: the drive to maturity

This is characterised by the spread of prosperity and economic development to the remainder of the country, and it also sees economic diversity occurring and the development of the tertiary sector with health, education and welfare facilities improving to the benefit of the population.

Instead of being the source of primary produce, these raw materials can be used to further develop the secondary industrial base. New technological and industrial processes not present in the take-off stage appear within the economy. Spain is considered to be at this stage of development, as the industrial foundation has developed to include the car and aviation industries as well as high-tech industrial parks. The multiplier effect, whereby successful industrial enterprises attract other forms of economic development thus creating further employment, services and wealth, can be seen to have assisted Spain's growth and development, as was the effect of joining the European Union (EU) in 1978. Further effects have seen Spain's transport infrastructure develop with road, rail and air links all now being highly developed. This has enabled the peripheral areas of the country to gain from the new prosperity. This has also been accompanied by a corresponding rise in the country's urban population and a fall in birth rate, all considered signs of a drive towards maturity.

e) Stage 5: the age of high mass consumption

This is dominated by expansion of the tertiary service sector and the decline in the manufacturing sector as automation supersedes manual labour. The consumption of consumer goods such as electrical

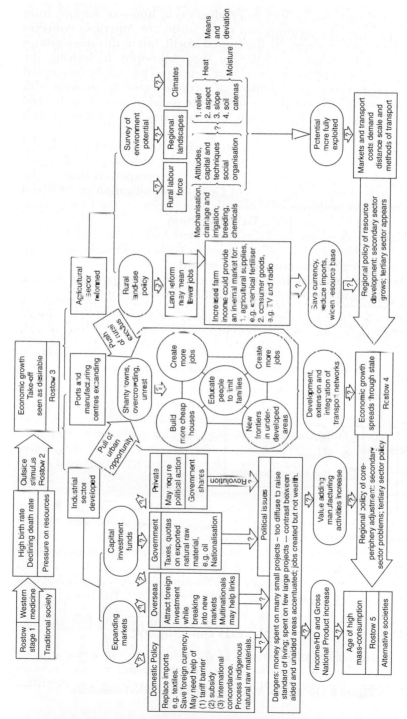

Figure 5 The Rostow model developed. *Source:* Horsfall (1982).

items and cars may be met by production at home or import, from abroad. Mass consumption and materialism are American concepts that have spread across the developed world, but at this stage groups within society begin to challenge the philosophy underlying the growth and a social 'conscience' may develop. This may be concerned with the environmental impact of previous growth, as seen with organisations such as Greenpeace, or the social impact of unchecked growth on groups within society that have not benefited from such gains, such as the elderly.

The USA was the first nation to reach this stage, and the UK and Germany are now also characteristic of the fifth stage of Rostow's model. Germany, however, is currently less prosperous than in the late 1980s owing to the effects of German reunification in 1990 and the resultant problems of economic integration of the former communist East Germany into a Western nation.

Figure 5 develops the ideas postulated by Rostow, and Figure 6 indicates when the take-off occurred in certain countries and when the age of high mass consumption started.

Rostow's work can, however, be criticised on the grounds that it is descriptive with no detailed explanation as to why. Linkages between each stage are not clearly explained and historical evidence suggests that the sequence is not necessarily followed. It is based on observations from western Europe and North America where the free market dominates and only looks at whole nations without looking at variations in development within the nations discussed.

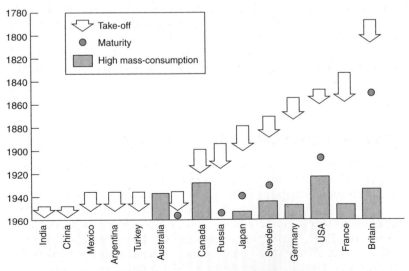

Figure 6 The Rostow model applied to selected countries.

Rostow's stages of economic growth are a good starting point for investigating MEDC and LEDC nations.

A subdivision of the tertiary sector, only categorised from the late 1960s onwards, is quaternary industry that incorporates financial and insurance activities. Research and development is also significant in this sector of industry with high inputs of scientific investigation and innovation leading to the production of technologically advanced products devised by a highly qualified workforce. Media activities and government agencies, with the role of producing workable ideas for application in society, come under the realm of quaternary industry, although a new term, quinary, is being applied to classify this area of activity.

4 An Investigation of Specific Primary Activities

Primary activities occupy the greatest amount of the Earth's surface, with farming and forestry being the two dominant activities.

a) Forestry

Large swathes of the Earth's surface were covered naturally by trees. The hardwoods of the tropics were matched by the hard- and soft-woods of the northern latitudes, in between which lay vegetation of lower height, size and density all associated with the climatic conditions that dominated a location. These natural forests have suffered mixed fortunes. The human colonisation and settlement of North America, Europe and Asia all saw the felling of the forests to create land for farming and settlement over the past 2000 years, while the tropical rainforest of the equatorial latitudes have only recently been felled in a similar manner.

Worldwide concern now exists over the fate of the world's forests as their role in global climate is now recognised, as is their vast biodiversity.

CASE STUDY: FORESTRY IN THE UK

The Forestry Commission directly harvests 5 million cubic metres of round timber, timber stripped of leaves and branches, annually accounting for 60% of the UK's softwood production.

In Wales, forest covers 14% of the land area. Of this, 70% is coniferous forest, but to ensure greater biodiversity, between the years of 1999 and 2000, planting of broad-leaved species increased to 25% of the total planting. Forests now cover 84,000 hectares of land in Wales, adding £400 million to the national

economy and securing 4000 jobs. Further expansion in employment is expected as softwood forests planted in the 1960s are
becoming mature enough for cutting and thus bringing about
an economic return on the investment of the past. By 2015,
Wales is expected to produce 1.9 million tonnes of timber
compared to the 1.4 million tonnes produced at the turn of the
century. Forests are a sustainable resource and, managed

Figure 7 Forest cover in the UK
(the darker shading represents more forestation)

properly, the UK should be able to produce more of its national requirement.

Presently, the UK produces 7 million tonnes of timber per annum from its forests, which is only 15% of the country's requirement. Investment of £1.6 billion in sawmills, paper and board mills over the past 15 years will be enhanced by a further £2 billion investment over the next 20 years to accommodate the expansion of the forest industry. None of the cut product is wasted. The tops of trees, branches and smaller trees cut when forests are thinned are used for paper and cardboard. Strandboard, chipboard and fibreboard, where fibres of timber from spruce and pine trees are glued together, are used for flooring and furniture. The trunks are used for flooring, furniture, planks and boards, while the bark is used as chippings and mulch for use in gardens.

Employment in forestry

The Forestry Commission directly employs 3900 people and together with private landowners 99,500 people are in full-time employment in forestry in the UK, with 50% employed in England, 36% in Scotland and 14% in Wales. Of these, 43% work in the forests themselves while the remainder are employed in primary wood processing, harvesting and haulage, maintenance and non-forest work including research and office administration. Employment has fallen 15% since 1993–1994, indicative of rationalisation and increased mechanisation common in all industrial sectors in the UK.

The UK's timber requirement

In total, the UK uses 42.5 million tonnes of imported timber and timber products at an annual cost of £8 billion. These are both hardwood and softwood from the northern forests of Russia, Canada and Scandinavia, while tropical hardwoods are imported from legally licensed forests of South-east Asia and South America. It is in such regions of the LEDCs that forest exploitation is becoming a global concern. Unlike the northern temperate forests dominated by stands of one or two species of trees, tropical forests contain many hundreds of tree species. To reach the economically important tree may lead to the cutting down of many other trees to get to what is required. These are then either burnt or left to rot and decay. Exploitation and a lack of replanting the extracted forest lead to the destruction of forest habitat, which has both local and global effects.

CASE STUDY: FORESTRY IN ZIMBABWE

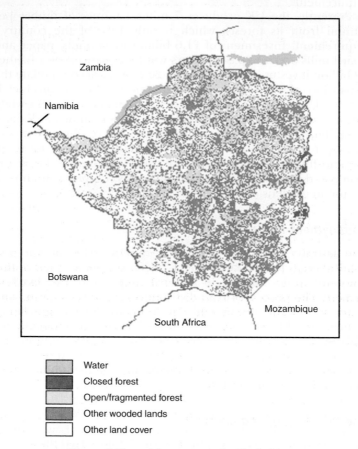

Figure 8 Map of forest cover in Zimbabwe.

The tropical savannah climate of Zimbabwe has 22% of forest cover, 31% of woodland and 13% of bushland. Subtropical woodland occurs in the east of the country with plantations of eucalyptus and acacia, while mopane occupies the northern Zambezi and Limpopo valleys in the south.

Like Britain, a Forestry Commission exists. Its role is somewhat different to the UK counterpart in that it tries to ensure forestry production while ensuring biodiversity conservation and it also promotes forestry against rural development. In common is the promotion of public service provision.

In Zimbabwe forestry is in decline as Figure 9 implies.

Land area	38,685,000 hectares
Forest cover (2000)	19,040,000 hectares
Forest cover change (1990–2000)	320,000 hectares
	−1.54% per annum

Figure 9 Forest cover in Zimbabwe.

Zimbabwean forests are primarily used for fuel and supply 75% of the country's domestic needs. This is the major reason behind the decline in the amount of forest. Even though the Forestry Commission does have a seed laboratory to assist with community seeding projects, increased clearing of forest for farming on the poor-quality communal lands that contain the highest density of population means its work is of little consequence. These communal lands occupy 42% of the country and are home to 75% of Zimbabwe's population.

The best land and 30% of the forests were in the hands of commercial farmers, but the country's current economic and political crises have led to war veterans taking over much of this privately owned land and redistributing it to the landless farmers. The process of land distribution is not considering any policy supporting sustainable development. Forests are being cleared for farmland and, in the case of fuel wood extraction, no replanting is taking place. This aids erosion and gullying in the wet season and the land becoming barren.

A side-effect of this is an 80% decline in the wildlife that roamed the forests and bushlands in a country once renowned for the amount and variety of animals it contained on ranches and in its national parks. Even Hwange National Park, adjacent to Victoria Falls in the west of the country, does not have enough money to fuel the pumps to fill the waterholes on which wildlife is so dependent. Poaching throughout the country is also having a devastating effect on the wildlife. For instance, instead of animals such as kudu being shot for game meat to sell to tourist restaurants (a major source of foreign exchange until the late 1990s), the snared animals are often left to die.

These actions have seen the demise of the Communal Areas Management Programme for Indigenous Resources (Campfire) introduced by the government in 1984, through which local communities received money earned from game-hunting fees and selling hides and meat. Cash payments for crop losses caused by wild animals were also paid. The protection this offered has now gone and 2 million Zimbabweans have lost a source of income, and the forest and wildlife have less protection.

Those forest industries still operating produce sawn timber, panels, pulp and paper mainly from plantation-grown wood. Recent trade has seen imports rather than exports dominating, with US$32.4 million worth of paper, paperboard and pulp being imported and only US$9.4 million worth of sawnwood, in particular, being exported in 2001.

The UK and Zimbabwe provide contrasting ways of managing a primary product, and show how success and failure are dependent on planning, political will and economic considerations.

b) Coal mining

Another primary source of fuel, and perhaps the most important mineral extracted on a global scale, is coal.

CASE STUDY: THE BRITISH COAL INDUSTRY

British coalfields

The development and exploiting of Britain's coal change led to a shift in the location of industry from forest to coalfield sites. The UK is fortunate to have vast quantities of coal, originating in the Carboniferous rocks, which also contain supplies of blackband iron ore and limestone. Coalfields stretch from the central valley of Scotland down through the north and north-east of England to the Midlands and South Wales. During the nineteenth century and up to the 1960s, coal was Britain's prime source of power generation and the foundation of the country's industrial might. The importance of the industry remained fundamental to governments of all political ideologies.

The 1960s to 1980s

Throughout this period, coal was a sensitive issue for government. Imports were banned in the 1960s and pressure was applied to industry to remain with coal as a power source, even with the new sources of power generation becoming increasingly available, especially cheap oil imports from the Middle East. The subsequent oil crisis of 1973–1974, which saw the price of imported oil increase, led to energy shortages within the country and the introduction of a 3-day working week to ration energy supplies. This crisis provided an opportunity for coal to regain its dominance as an energy source. The opening of a new pit in Selby, north Yorkshire, which had the largest untapped reserves in Europe, and one in the Vale of Belvoir in

Leicestershire, saw 10 million and 7.2 million tonnes of coal mined each year, respectively, and heralded a new era for the industry. However, output was low at 2.44 tonnes per miner, making it less competitive compared with foreign imports.

The 1980s onwards

The following 10 years saw a decline in the coal industry, with a reduction from 172 pits in 1984 employing 174,000 workers to 15 pits in 1994. Thirty-one pits were closed in 1992 with compulsory redundancies totalling 30,000 miners. The National Coal Board (NCB) was de-nationalised by the then Conservative government in 1994 and RJB Mining became the country's leading producer of coal employing less than 19,000 miners in fewer than 20 pits or collieries. However, output per miner had increased to 8.76 tonnes per miner by the time RJB took over, thus making it a more efficient industry than before.

Coal requirements in the twenty-first century

UK Coal now runs the largest coal producer in the country and saw production rise from 19.1 to 19.6 million tonnes of coal between the years 2000 and 2001. There was a 10% increase in coal use in thermal power stations in 2001 and an order book of 50 million tonnes up to 2010. The industry still receives aid from government, but only to a maximum of £75 million, of which £53.3 million and £21.7 million were received in 2000 and 2001, respectively, to counter losses that in 2001 were £26.5 million.

In 2001, the UK consumed 65 million tonnes of coal of which 32 million was mined in the country; 17 million tonnes came from deep-shaft mines and 15 million tonnes were from opencast mining. The remaining 33 million tonnes were imports that came from South Africa, Australia, Poland, Russia, the USA, Indonesia and China. Fifty-one million tonnes were used in thermal power stations, with the balance being used in steel and other industrial and domestic uses. Of fossil fuels, coal is the cheapest generator of electricity costing 1.5p per kilowatt/hour compared to 2.2p at gas-fired power stations. In the future, environmental considerations rather then economic factors will cap the amount of coal burned.

The EU-wide combustive plant directive comes into effect in 2008 and will affect non-modified power stations limiting carbon dioxide emissions. This may affect production, but it is anticipated that the UK will continue to produce 30 million tonnes of coal per annum for the foreseeable future.

Mines will close as coal reserves become exhausted, but communities will continue to exist in these locations as new industrial opportunities take over and ensure future economic activity and development.

CASE STUDY: THE BULGARIAN COAL INDUSTRY

Bulgaria is a former Eastern bloc country on the verge of joining the EU as a new European democracy in around 2007. It has 22 heavily subsidised mines, nine of which are opencast and 13 of which are shaft mines. In 2001, 26.6 million tonnes of coal were mined. Production is declining due to reduced electricity consumption in the country linked to a fall in purchasing power of the Bulgarian people and low labour productivity. Research by Markov (2001) highlighted the limited disposable income of the Bulgarian people and the shortage of money as a problem for the mining industry leading to decapitalisation of the economy and the country becoming unattractive to potential foreign investors.

Insufficient investment has prevented new mining units from being developed. Outdated and worn-out equipment to extract or process the mined coal has not been replaced and has led to the decline in output shown in Figure 10.

Year	Coal production (tonnes)
1989	35,801,000
1990	35,060,000
1992	31,423,000
1994	29,782,000
1996	32,363,000
1998	30,962,000
2000	26,278,000
2001	26,586,000

Figure 10 Annual coal production in Bulgaria, 1989–2001.

Only one mine, Mini Maritsa East EAD, is financially stable. This mine supplies state thermal power stations and is essential as dependence on ageing nuclear power stations is reduced.

Bulgaria's energy strategy proposes remedial action – the modernisation of heavy machinery, the introduction of new equipment to improve productivity, as well as the protection and restoration of the natural environment. With shaft mining, more competitive methods of production and efficient mining techniques will be encouraged by restructuring and privatising the industry, thus allowing US$400–500 million of outside investment to be put in place.

The World Bank expects a 3.5–4.5% growth in gross domestic product (GDP) to be reached through secured production of primary industry, of which 45% would be based on coal by 2010 reducing to 43.7% by 2015. This would see an annual production of 30–34 million tonnes of coal up to 2015. To this end the Ministry of the Environment and Waters will introduce a goal-oriented budget, secure within human and financial resources in line with Bulgaria's new capitalist approach to the economy.

Questions

1. Give detailed definitions for the terms primary, secondary, tertiary, quaternary and quinary industry.
2. Describe how Rostow's model can be used to explain economic development.
3. Using the triangular graph (Figure 2), rank the countries according to their level of development.
4. What does a high level of employment in: a) primary and b) tertiary industry indicate about the level of development in a country?

2 Industrial Location Theory and the Traditional Manufacturing Industries

1 Industrial Location

Each industrial unit, whether it employs one person or is a large factory employing thousands of people, has its own unique site. The land the unit occupies is called its site and when choosing that site the entrepreneur, or executive responsible for such decisions, will take into account many factors. These will depend on the nature of the

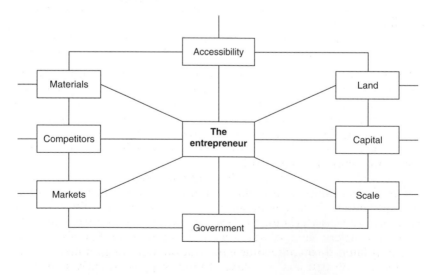

Figure 11 The factors an entrepreneur will consider when locating an economic activity.

business activity and also the time frame as to when the business was being established.

Traditional industrial location theory would have had the early entrepreneurs looking at the availability of raw materials and the market for the product in particular. It is only in more recent times that other factors have become important influences on industrial location.

These additional factors to be taken into account are both physical and human. Physical factors include the land itself, the raw materials required and the accessibility that the location offers. Human factors include the market being served, the labour force, competitors for the product, availability of capital and the scale of the operation. Furthermore, government policy is also a factor that the entrepreneur will need to consider. The linkages between these elements are shown in Figure 11, which is a systems diagram.

a) Materials

All industries require inputs in the form of materials, as these are the foundation of industrial activity. Their availability can determine the location of industry. The extraction of raw materials from the ground brings about the establishment of quarrying and mining centres that may also become centres of other industrial activity.

For example:

- A company producing concrete requires inputs of sand, gravel, water and cement (a mixture of chalk and clay).
- Sand and gravel are usually located together owing to their similar geological origins.
- Chalk and clay may be located together but at a different site.
- Water is a ubiquitous material, which is found in all locations.

The entrepreneur, therefore, has to decide where to locate. There are choices available for the cement factory either near the source of the sand and gravel or near to the source of the chalk and clay. In either case, there will be a need to transport the other raw materials to the site of the factory and it may transpire that the best location may be at an intermediate site to where all materials can be assembled for manufacture at a lower price.

Primary activities such as mining and quarrying can only take place where the raw material is available, but it also needs to be accessible to exploitation. This may depend on the level of technology at the time the enterprise is being established. Mining for coal was originally limited to where the coal seams were exposed at the Earth's surface and only later, through technical innovation, were deep mines able to exploit the unseen coal reserves. Mining coal, as an early source of power, became a significant industry in the nineteenth century and was the driving force behind the industrial revolution, and if the sites of coal mines also had reserves of black-band iron ore they were likely to become centres of secondary industry.

b) Labour force

To exploit the raw materials a labour force is also needed and the availability of this may vary. This requirement may require a small number of skilled people or large numbers of unskilled people, but they may not live near the site of the raw material due to it being a newly discovered resource. Unless people can be encouraged to migrate to the site to facilitate exploitation, it may not be possible to use the resource. This may therefore depend on the incentives given and the economic outlook at the time. The former Soviet Union gave incentives for people to migrate to the harsh climatic region of Siberia to exploit the vast resources that this isolated area of the country harboured. In reality, even with such incentives, many people found coping with the extreme cold and isolation of Siberian mines too much and returned to the cities of western Russia.

c) Markets

The market for the product being produced may be local, national or international, or a combination of all three. The product may also be lightweight or bulky and it may be perishable. These factors need to be considered by the entrepreneur when choosing where to locate, as the aim of the industrial activity is to make a profit. The bulkier or more perishable the product, the closer to the market the industrial site needs to be.

d) The entrepreneur

The entrepreneur who also has a personal location will need to evaluate where best to establish the site to satisfy the market and ensure that the enterprise is viable and profitable. The significance of the birthplace of the entrepreneur needs to be stressed as this can influence decisions made regarding the location of an enterprise. The entrepreneur will assess this home environment in terms of the possibilities for industrial enterprise taking into account the resources an area possesses or the needs of the community it serves. Cadbury's at Bourneville, Birmingham and Morris Cars, the forebears of BMW's Mini car factory, at Cowley, Oxford are examples of companies located close to the birthplace of their founders.

However, if the entrepreneur is producing ready-made concrete, a location close to the market for the concrete would be essential. As deliveries by mixer truck need to take place within 90 minutes of mixing, before the concrete sets inside the mixer, a location within such a time-distance from the market is essential. This would be the only way in which the market could be effectively served and a raw material-based location would be chosen.

If the entrepreneur was a baker serving a local area, a location close to the market would be chosen to ensure prompt delivery of the fresh produce. This is a market-based location. Other more sophisticated products will serve a larger area and the size of the market area is, in part, determined by the kind of product or commodity being produced, and, in part, by the cost of distributing that commodity to the customer. These two considerations play a significant role in the decision-making process of the entrepreneur when deciding where to locate an industrial enterprise. Other considerations are:

- competition
- accessibility
- land
- government
- scale of production.

e) Competition

In a free-market economy, competition can come from other companies seeking to take a share of the identified market at a particular location. In the twenty-first century, competitors may be local, national or international depending on the product being made available for sale. The competition may come from the development of new products that may bring about a reduction in the sale of a local, established product.

f) Accessibility

Access can take many forms. Apart from raw materials, the labour force or the market, it may also be concerned with the ease in which component parts, housing, banking and insurance activities can be accessed. Environmental considerations and even a golf course may be considered. It can also refer to the proximity of an international airport or docks for the import or export of the product. However, deliberate inaccessibility may be the preferred site for defence industries and research and development institutions owing to the nature of their work. However, as A.G. Hoare found in a study of 30,000 companies of varied sizes in Greater London, access to materials, market and labour are the most significant factors of industrial location.

g) Land

The land required by an industrial unit may vary in size, depending on the nature of the industry. Small sites are easier to obtain, but larger areas to accommodate a car assembly plant, for example, are much harder to locate and develop. The availability of a suitable site, bearing in mind the considerations the entrepreneur has already identified, may be difficult. The larger the site, the more problematical it may become to find the ideal or optimal location. The cost of land and premises may also vary depending on location, e.g. land and premises in south-east England have much higher costs than elsewhere in the UK. If the product being produced can be as easily produced outside of the market area or country, in an area where land is cheaper, an entrepreneur may choose to locate the plant there.

h) Least-cost location

Least-cost location and profit maximisation may not be the prime consideration for an entrepreneur. The entrepreneur may be satisfied with a suboptimal location and the profit that can be gained from such a location.

The reconciling of all of these influences and the decisions made by each entrepreneur produce the landscape as seen today.

i) Capital

The availability of capital to establish an enterprise can also affect the decision-making process, e.g. the entrepreneur will need to decide on the location best suited to needs to ensure that the capital is being wisely invested in the chosen location.

j) Government

Government intervention may play a role in the decisions the entrepreneur makes. The political philosophy can determine the role of government and the way in which an industrialist/entrepreneur may react to it. There exists a continuum from socialist to free-market principles that operates around the world. In the UK the free market and mixed economy regulations dominate. while in France land ownership is more in the hands of the state and hence decisions about industrial development and location of industry. In the UK, incentives to locate in particular areas, such as Enterprise Zones, may be given and land may be made more available for development in certain areas. Similarly, subsidies and grants may also be available to ensure what the government feels is a fairer distribution of industry. It is beyond total free enterprise or the *laissez faire* view of an economy based on choice by the entrepreneur to locate, but is a fact that has to be recognised in industry today.

Early industrial location theory was primarily concerned with the importance of labour, market and raw materials, and this is best postulated in the work of Alfred Weber.

2 Industrial Location Theory

a) Weber's least-cost location theory

Alfred Weber, a German economist, devised in 1909 the theory of industrial location. It was based on the concept of least-cost location that would enable profit maximisation.

Weber was concerned about the relationship between the locations of raw material, market and the significance of transportation between these. He looked at individual factories and in so doing he made a series of assumptions from which later generalisations could be made.

Weber's assumptions:

- An isotropic plane with ease of access in all directions.
- An uneven distribution of natural resources with concentrations at particular locations.
- Labour is at fixed locations with fixed wage rates, and is immobile and unlimited.
- Culture, climate and political systems are uniform.
- Markets are fixed in size and location on the plane.
- The entrepreneur will always minimise the costs of production.
- Perfect competition exists.
- There is no variation in the cost of land, structures, equipment or capital.
- There is a uniform transport system over the isotropic plane.

Weber used these assumptions to study the theoretical location of industry and later apply them to the real world.

Weber considered the situation as to whether an industry should be based at either the location of the raw material or at the market. He looked at the raw materials involved and whether there was either a weight gain or a weight loss in the production process. No gain or loss would mean that either site could be chosen, but if there was a weight loss during production this would favour a raw material location, whereas a weight increase would favour a market location. The gain or loss of weight directly affects the bulk of final product to be transported and hence transport costs. As a profit maximiser, the entrepreneur would always favour the location of the factory at the lowest level of transport cost.

This led Weber to develop the materials index that states:

$$\text{Materials index} = \frac{\text{Total weight of the located raw materials used to manufacture the product}}{\text{Total weight of the finished product}} .$$

Located raw materials are those raw materials that are only found in certain places, such as iron ore and coal, and ubiquitous materials are those raw materials that are generally found everywhere, e.g. water. The above equation does not include ubiquitous materials.

The resultant figure will either be below, at or above 1. A value of 1 indicates no gain or loss of weight in production. Less than 1 indicates the product has gained weight in production and this favours a market location. Greater than 1 indicates that weight is lost in production and this favours a raw material location.

For example, if 2 tonnes of material *a* and 2 tonnes of material *b* are needed to produce 1 tonne of a final product, weight loss occurs in the manufacturing process and a raw material location would be advisable as transport costs would be reduced. If 1 tonne of material *a* and 1 tonne of material *b* produced 2 tonnes of the final product, a market-based location would be attractive, especially if the product was perishable, e.g. bread.

Such profit maximisation may be affected, however, if labour costs were cheaper at a different location and outweighed the additional costs of transportation to that point. This Weber illustrated by isodapanes, or cost contours, which are lines of equal transport cost per unit of production from site P. If cheap labour was available at sites L_1 and L_2, which would reduce the overall cost of production by 25p per item, the entrepreneur would save more on labour than expended on transport if the location was within the critical 25p isodapane. As shown in Figure 13, location L_1 is within the critical isodapane while L_2 lies outside it. Therefore, L_1 would be an advantageous site while L_2 would not.

Weber recognised the importance of labour costs as he saw improvements in efficiency of transport develop. He also looked at agglomeration. This is concerned with the advantages that may be gained by companies locating together.

Again, a critical value isodapane is used at each site under consideration and if they overlap, and transport costs are not increased beyond the value of the isodapane, then a new location would be considered to be profitable.

Weber's work has been criticised because of the assumption that raw materials and markets were at fixed points only. His assumptions of 'economic man' operating under perfect competition are also criticised as is the static nature of the model which fails to take into

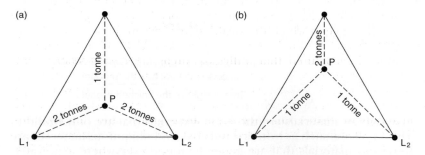

Figure 12 Weber's locational triangle. *Source*: Knowles and Wareing (1985).

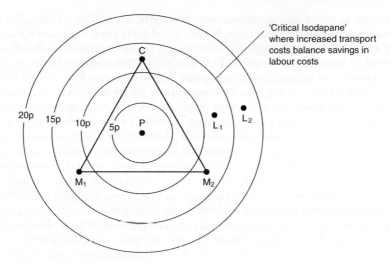

'Critical Isodapane' where increased transport costs balance savings in labour costs

Figure 13 The affect of labour and transport costs on location. *Source*: Knowles and Wareing (1985).

account changes in supply of raw materials or the demand for the product in the marketplace. Industries are not simply processors, as the final product of one industry often becomes the raw material of another product higher up the scale of technical sophistication.

Weber's model is also dominated by transport costs directly proportional to distance, which is not necessarily the case, and an inflexible rather than a mobile labour market.

b) Myrdal's model of cumulative causation

The ideas put forward by Weber were challenged by the Swedish economist Gunnar Myrdal in 1957 and his view of industrial location had evolved looking at less-developed economies. He saw a three-stage model where, in the first stage, a pre-industrial situation existed where regional differences were minimal and was followed by a second stage of rapid growth that occurred in certain regions only. This led to increasing economic divergence between regions and led to a third stage in which the wealth generated in these most affluent regions filtered through and spread to other areas of the country.

Myrdal recognised that individual areas are either economically advantaged or disadvantaged. Those advantaged areas will generate investment and will be characterised by economic growth. This growth, once initiated, will generate flows of capital, labour and raw materials to further enhance these areas growth and development. This expansion can be referred to as the cumulative causation process (see Figure 14), which has a negative, or backwash, affect on the less-developed regions as skilled labour and locally generated capital is drawn to the more economically attractive area. The economies of scale can be seen to make the locally produced goods and services less competitive even though they lie within the hinterland of the dominant 'resource rich' region. This leads to a polarisation between the resource and industry, rich and poor areas respectively resulting from a growth pole seeing the concentration of industry in one region only. Time, however, can create problems where such agglomerations of industry exist, with an increasing inability of the services in an area to meet the demands of the industrial complex that has evolved. At this stage the concept of cumulative causation may lead to the development and creation of new centres of self-sustained economic growth in the hinterland where raw materials or other economic advantages are seen to operate.

The theories rely on 'economic man' being rational and in receipt of all relevant knowledge. Reality, however, is different to this and industrial location depends on people as optimisers. In this case the entrepreneur is not in receipt of all information and may choose a suboptimal site. It may still be profitable, but not as Weber or Myrdal would have expected.

With these location theories in mind, the studies below look at the growth of the traditional industries of textiles and steel in England and the USA, respectively.

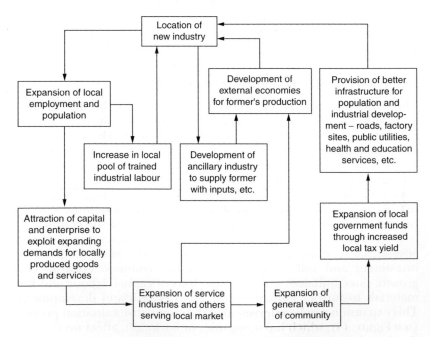

Figure 14 Myrdal's theory of cumulative causation.

3 Early Industrial Location

a) The English textile industry

The eighteenth and nineteenth centuries saw massive expansion of industry, but not in the traditional locations of south and east England, but in the north of England where the inventions of the 'industrial revolution' could be best implemented.

CASE STUDY: THE LANCASHIRE TRIANGLE – TEXTILE INDUSTRY

The greatest concentration of textile factories was found in the Lancashire triangle containing Oldham, Bolton and Manchester. This area had an experienced workforce, new machinery, the necessary moist climate and the appropriate source of power, first water and then coal. Manchester, in particular, had been transformed from a market town to a major city with the construction

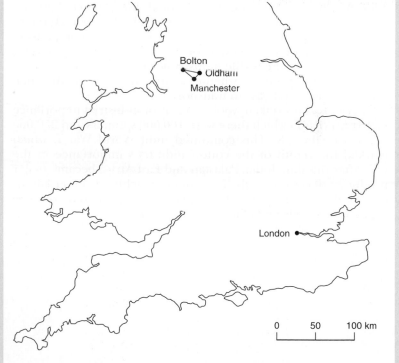

Figure 15 Map of the Lancashire triangle.

of the Duke of Bridgewater Canal in 1761 to import cheap coal. Between 1771 and 1831 there was a sixfold increase in population as people migrated to the city and by 1851 it housed a population of 455,000. The cotton that brought about this growth was imported via Liverpool, since Manchester is a land-locked city. Tolls and harbour dues at the port of Liverpool reduced the profits of Manchester's merchants and a canal linking the Mersey estuary with Manchester was proposed. This was to become the Manchester Ship Canal which, following 5 years of construction, was opened in 1894 at a cost of £15 million. Such was the importance of cotton to Manchester that such a capital outlay was made.

However, not all of Lancashire's cotton industry benefited in a way similar to Manchester. Companies could not automatically take on any new invention as soon as it became available, as money had been invested in a new plant and machinery. Consequently, although Watt's steam engine was available after 1789, water power was still in use in the textile industry in the middle of the 1800s.

However, by 1840, 87% of the mills were steam powered and located on coalfields. This was essential in keeping down costs as a location 8 miles from a coalfield would, for instance, have led to a doubling in the price of coal, as it is a bulky raw material to transport and, in turn, makes production less competitive. Consequently, it was the power source that became as important to the location of the textile industry as the raw material used to make the final product, whether cotton or wool, as the final product is light and easy to transport.

Cotton had overtaken wool in terms of industrial importance by 1812, a year in which there were 100,000 spinners and 250,000 weavers in the UK. This continued until World War I, which marked the zenith of the cotton industry's importance in the UK. After this time India, Pakistan and East Africa became major producers of cotton goods. Their newer machinery, lower labour costs and cheap transportation meant that they could compete with cotton goods manufactured in the UK.

The Lancashire triangle is still a producer of textiles although its importance is less and artificial fibres rather than wool and cotton are used.

b) The steel industry

> ## CASE STUDY: BIRMINGHAM AND
> ## THE IMPORTANCE OF THE METAL INDUSTRY
> ## TO ITS DEVELOPMENT
>
> ### Nineteenth-century developments
>
> It was the first 60 years of the nineteenth century that brought
> about the emergence of Birmingham as the industrial and com-
> mercial centre of the English Midlands. Most products were pro-
> duced by 'small masters' in workshops contained in their own
> homes or attached to them. Arms manufacture was the first to
> develop, but by the middle of the century the jewellery trade
> employed 9000 people making both high-quality products and
> those for the less affluent market. A further 6000 were employed
> in button manufacture, a figure that had remained almost con-
> stant throughout the century. Brass goods were also a staple
> Birmingham industry with iron foundries producing items for
> such diverse markets as the navy and plumbing. Railway devel-
> opment increased the demand for brass tubes and the size of the
> manufacturing unit developed from being home based to larger
> purpose-built units that also produced a variety of other metal
> goods.
>
> By the 1850s Birmingham became a focus of a network of rail-
> ways and this provided greater accessibility that enabled the
> development of retail and commercial activities in the town
> centre. The railways served the 'Black Country' as well as
> Birmingham itself. However, the industrial developments that
> had evolved had taken most of the available space within the cen-
> tral area and new developments began to grow, initially on the
> eastern fringe but later in all directions following road, canal and
> rail routes.
>
> Companies such as BSA (Birmingham Small Arms) in 1861
> and Cadbury's in 1879 are examples of companies established to
> make such diverse products as guns and chocolate in purpose-
> built factories.
>
> ### Into the twentieth century
>
> The skilled artisans associated with Birmingham were a magnet
> for sustained growth and this continued throughout the nine-
> teenth century and saw further expansion on the fringe of what
> was to become a city in 1889. Although Birmingham missed out
> on the electrical engineering industry that began to evolve from
> the 1880s and was located in London and the south-east, where
> an agglomeration of demand and skilled labour existed, it was

the emergence of the motor vehicle in the twentieth century that put Birmingham on the map. Birmingham was in a good position to exploit this new invention as its experience in metal working enabled the development of a motor and component industry. The Longbridge works in Northfield, developed on a 2.5-acre derelict site of a former printing works in 1906, now covers in excess of 100 acres and currently produces MG and Rover cars.

After World War I, Birmingham became the home of British car manufacture, competing with US makers General Motors (Vauxhall), established at Luton in 1928, and Ford, established in Dagenham, Essex in 1932. Austin, Morris and Singer were all makes of British car at this time. Component manufacturers such as Dunlop tyres, Lucas electrical goods, Serck radiators and Triplex glass all moved or expanded their operation in Birmingham in the 1920s and 1930s.

However, as the 1951 census noted, 72% of those employed in manufacturing in the city were directly or indirectly involved in motor manufacture. This continued throughout the 1960s, a time of considerable industrial unrest and hostility to change among the workforce. However, such specialisation and dependence on one industry was to prove economically disastrous during the 1970s when decline in the car industry came about largely because of foreign competition.

Birmingham lost the advantages of early industrialisation in the 1960s and 1970s. Outdated plant and a lack of investment, due in part to the city's high wages and rents, had discouraged many companies from locating in Birmingham. It made such a move uneconomic and uncompetitive while, at the same time, the newly emerging industrial nations of the Far East were more competitive. Companies such as Datsun from Japan began to eat into the traditional home market of companies such as British Leyland. This economic slowdown and the recession of the 1970s and early 1980s, culminating in unemployment levels of 14.9% in 1982, laid waste to the earlier prosperity of the city. Birmingham later re-invented itself as a post-industrial city (see Chapter 3).

CASE STUDY: INDUSTRIAL DEVELOPMENT AND DECLINE IN NORTH-EAST USA

Industrialisation

The zone of industrialisation extended from Buffalo in upstate New York north-westwards through Pennsylvania, including Pittsburgh, and northwards through Ohio including Cleveland.

It also encompassed southern Michigan, especially Detroit, and the Great Lake cities of Chicago and Milwaukee. Eight states are partially encompassed within this industrial belt, as are the southerly shores of three of the five Great Lakes (see Figure 16).

The area possessed the necessary raw materials, ease of transportation and a rapidly expanding workforce. The industrial ingredients of heavy industry were present with coal being mined in the Allegheny Mountains of Pennsylvania and iron ore being extracted on the western shores of Lake Superior. Limestone was also available and all three raw materials could be transported via the Great Lakes or the developing rail network to the industrial centres that were developing within the region. The opening of the Erie Canal in upstate New York also enabled the lakes to be linked with the Atlantic Ocean via the River Hudson, avoiding the St Lawrence River that freezes over in winter.

The ensuing industrialisation of the area in the late nineteenth century saw the growth of cities such as Duluth, Chicago, Cleveland, Pittsburgh and Buffalo.

The steel and car industries

It was Henry Ford who formed the Ford Motor Company in 1903, as one of 15 car manufacturers in Michigan. In 1913 he introduced the moving assembly line at his Highland Park plant enabling workers to perform one task only and thus improve efficiency and production level. This revolutionised the industry, making vehicles more affordable and therefore more were produced to meet the increased demand. Fifteen million Model 'T' cars were produced before 1939. Other companies such as Buick and Oldsmobile followed suit and enabled the area to prosper. Ford's prosperity continued during and after World War II. Ford is now a multinational company owning the Lincoln and Mercury car brands in the USA, Mazda in Japan, Jaguar, Land Rover and Aston Martin in the UK, and Volvo in Sweden.

Chicago, with a population of 7.5 million, was also a steelmaking centre and, owing to its strategic position, is a major transport node on the banks of Lake Michigan. It too became a centre of steel production complementing its earlier role as a transhipment centre for the grain and livestock from the Great Plains which led it to prominence in the nineteenth century. The chemical industry, plastics and oil refining were later additions to its domination of the north-eastern segment of the country. Using the concept of critical isodapanes as postulated by Weber, Chicago lies within the critical 80% isodapane centred on the prime location of New York, owing to the site's location being easily able to import the raw materials necessary for economic development.

Decline in these traditional industries has been matched by newer industries being attracted to the city due to its location and its sheer size. The late twentieth century sees tax concessions being given to high-tech companies that wish to locate in Chicago and thus reverse the decline in the industrial backbone of the city.

However, smaller cities were not as able to diversify. Gary, Indiana, is a city lying to the east of Chicago and in close proximity to Lake Michigan, and was established in 1906 by the United States Steel Corporation. It was named after its chief organiser. The site took advantage of being located on a navigable waterway between supplies of iron ore to the north and coal to the south. The Gary Land Company laid out a city for its workforce and its blast furnace was first fired in 1908 leading to steel production in 1909. The city was prosperous until the 1950s but Gary was essentially a one industry city and suffered from the overall decline in the area's steel industry and protracted labour disputes. The consequence of all this was a population decline of 7% between 1909 and 1998, and the decay and dereliction of its industrial fabric.

The 'rust belt'

Most employment in the manufacturing industries of the northern Midwest was typical of the 'old' blue-collar manual work. The workforce was also highly unionised. As these industries saw competition develop from the newly emerging economies of the Far East and plant was unable to compete with the new product, many companies abandoned their traditional centres of manufacture in the area. This led to dereliction of much of the industrial heartland of the Midwest with subsequently high levels of unemployment and the outward migration of large segments of the workforce to the 'sun belt' states of the south and west.

The term 'sun belt' came into use in the 1970s when the economic and political effects of the earlier migration were recognised. Newer industries, a less extreme climate and the possibility of living in a less polluted environment were keys to this migration. The southern states centred on Florida, Texas, Arizona and California have also attracted considerable economic inward investment since 1945, and by 1990 cities such as Los Angeles, Dallas and Houston were listed among the top 10 cities within the country. (California would, in its own right, be the fifth largest economy in the world after the USA, Japan, Germany and the UK, and has been a magnet for economic growth.) This was at the expense of cities such as Cleveland whose secondary industries had not been replaced by the newer tertiary industries that now dominate employment industry in such an economically developed nation.

Route 66, popularised by the music of rhythm and blues, is the highway linking Chicago with Los Angeles in California and was seen as the route out of this area of industrial decline. The term 'rust belt' is applied to the area owing to the decay and dereliction seen in many metropolitan areas as former factories fell into a state of decay and areas became abandoned. Some have re-emerged with newer industrial developments. Pittsburgh has become a centre for research and development, and has developed its role within the world of finance. However, others, such as Gary, Indiana, have not benefited from new initiatives aimed at countering the concept of decay as symbolised by the term 'rust belt'.

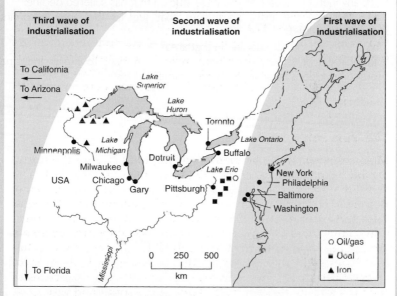

Figure 16 Map of the industrial north-east of the USA.

Industrial dynamism

The cotton industry and the steel industries of the UK and the USA are key examples of industry whose location was based upon the ease of access to raw materials. Once the initial advantages of location as postulated by economic geographers such as Weber and Myrdal had been lost due to advances in technology, new sources of raw materials and new markets, these areas went into decline and had to re-establish themselves, where possible, with a new industrial framework. This has been the challenge for cities such as Birmingham and Manchester in the UK, and Gary and Chicago in the USA.

Questions

1. Identify the differences between Weber's least-cost location theory and Myrdal's approach.
2. Describe the factors that an entrepreneur needs to consider when establishing an enterprise at a particular location. Which of these factors are not considered by Weber?
3. Explain the following terms:
   a) least-cost location
   b) isodapane
   c) materials index.
4. For any industrial area that you have studied that has suffered decline:
   a) Describe the initial advantages that led to this area becoming industrialised.
   b) Why did industry decline in that area?
5. To what extent is the Weber model relevant in trying to explain industrial location decisions taken at the present time?

3 Industrial Change and Behavioural Theories of Industrial Location

KEY WORDS

'Economic man': an entrepreneur in receipt of all relevant knowledge and able to apply it effectively
Losch's location theory: a theory of maximisation of revenue, which serves the largest market area
Smith's space cost curve: a graph that shows the relationship between costs and revenue
Pred's behavioural matrix: a theory where entrepreneurs are satisficers, not optimisers
Kirk's behavioural environment: a model that relates the physical environment to the behavioural environment
Footloose industries: industries not tied to a particular location

1 Introduction

The dynamic nature of manufacturing industry has had both a positive and a negative impact depending on the area of the world being considered.

The older, established areas of manufacture within Europe and North America experienced the negative effect of being in the first wave of industrial development with plant closures and the loss of employment as they became less competitive with the newly industrialised countries (NICs) of the Far East who benefited from investment in new technology that became available from the 1960s onwards and made industry more mobile or footloose. Once money has been invested in capital infrastructure within an industrial complex it needs to be utilised for many years before the initial costs of installation are met. However, as new manufacturing techniques are constantly being developed, which are improvements on earlier technology, this earlier machinery looses its competitive advantage. Consequently, the quality of the product being made by those companies who have invested in the new technology outweighs that of its competitors who go into relative decline. Later, even newer technology outperforms the second-generation improvements and so this dynamism continues and brings into play a cyclical process of industrial advantage and disadvantage.

2 Technological Change from the 1960s

Until the 1960s the world's manufactured goods tended to be produced in the USA or western Europe who jointly accounted for 90% of the world's output. This figure has since fallen to below 80%, while the NICs have seen a fourfold increase in manufacturing output. It is the nations of the Far East, starting with Japan, that harnessed the new industrial processes and the success soon spread to the neighbouring countries of South Korea, Taiwan, Malaysia and, more recently, China.

With regard to textiles, UK dominance was broken as cheap labour, suitable climate and local supplies of cotton led to the expansion of the industry in India, Pakistan and other nations of the Far East. Transportation of lightweight products added little to the cost and so clothing became cheaper than that produced in the UK. Textiles are still the ninth largest sector of manufacturing within the UK with a turnover of £17.7 billion per annum, of which £5.1 billion is exported, but output is in decline with a 30% fall between 1997 and 2001. Exchange rate fluctuations and the loss of comparative advantage compared with the low-wage economies of the LEDCs, to which many UK retailers such as Marks and Spencer have turned, has affected and reduced production levels in the home market.

The result of this is to reduce the number of people employed in this industry sector from 603,000 in 1988 to 264,000 in 2001. In August 2001, the industry was seeing an average of 2000 job losses a month.

Similarly, the automotive industries of both the UK and the USA suffered similar reductions in capacity from the 1960s onwards as a result of competition from the NICs. The UK car industry, which suffered from underinvestment, saw its market share fall and the numbers employed in the industry decline. (See the section on Birmingham later in this chapter and the rust belt in Chapter 2.)

However, the UK still produces in excess of 1.5 million cars annually from a workforce of over 47,500, even though the majority of car manufacturers are today foreign owned.

Economic changes and modern technology have given rise to new models of industrial location to compete with the earlier works of geographers such as Weber.

3 Beyond Weber's Theory of Industrial Location

Theories about industrial location can be placed into one of four categories. The first is that put forward by Weber (see Chapter 2) whose primary concerns were based on cost factors. Three further schools of thought have developed in the past century:

- locational interdependence
- demand
- profit and human behaviour.

a) Hotelling's model of locational interdependence

In 1929 Harold Hotelling looked at the impacts of demand on location and the decisions of entrepreneurs. In his model, itself a simplification of reality, he used the example of two ice-cream sellers working on the same beach. The assumptions made by Hotelling in this model recognised:

- an even distribution of population
- two competing entrepreneurs (a duopoly) with equal costs of production and the ability to supply the whole market with an identical product
- an infinitely elastic demand
- uniform production costs
- an ability for the entrepreneur to move without incurring any cost.

Hotelling assumed that if a second ice-cream seller were to set up in competition with an already established business, the competitor would need to consider carefully where to locate to obtain maximum sales of his product. Location away from the established competition would draw some trade from the rival operation, but may not maximise market share (see Figure 17), but a move to a location closer to the other ice-cream seller would result in an increase in sales for the new entrepreneur (see Figure 18). Both people would try to oust the competition, but compromise would lead to a relocation of both sellers to points enabling each to serve 50% of the market. The addition of a third competitor would see a similar process happening until the beach was divided into three sectors each with a smaller market area.

The advantage of Hotelling's theory is that it recognises the importance of competition within the marketplace for a particular product.

The work of both Weber and Hotelling assumes that the entrepreneur, or decision maker, is an 'economic man', being in receipt of all the relevant knowledge and possessing the ability to use it effectively. This enables the decision maker to choose and operate from an optimum location where profit will be maximised.

Such 'economic men' with perfect knowledge are, in reality, unlikely to exist and decisions will be made by individuals or groups of entrepreneurs with less than perfect knowledge or ability. However, their function is still to choose a location where costs are met and profit can be made.

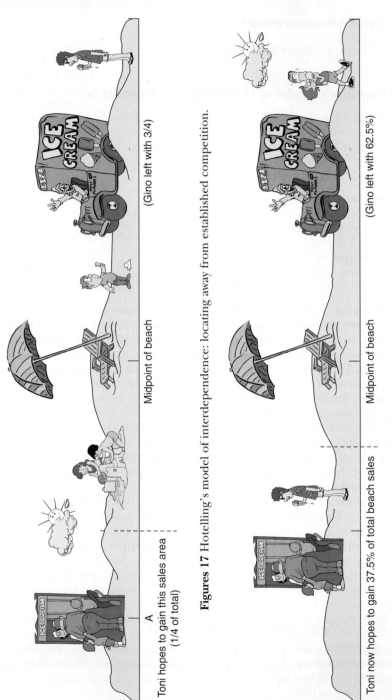

Figures 17 Hotelling's model of interdependence: locating away from established competition.

Figures 18 Hotelling's model of interdependence: locating closer to established competition.

b) Losch's location theory

Losch's general location theory was contained in a text first published in Germany in 1939 and translated into English in 1954, and it is considered to be the first attempt to devise a general location theory. Losch considered the optimum location to be the one that maximised revenue while serving the largest market area. His theory made certain assumptions:

- an isotropic plane
- an evenly distributed population
- costs of transport were proportional to distance
- costs of production were uniform
- uniform purchasing power.

Losch identified that with such assumptions the market area for an optimum location would be circular, with a reduction in demand occurring with increased distance from the point of production as costs of transport rise. Such circular market areas either leave some locations unserved or overlapping and he chose the shape closest to a circle, a hexagon, to avoid either situation.

However, as different products with different market areas will produce different sized hexagons, Losch recognised that a complex network of market areas will exist over a landscape. To accommodate this he rotated each hexagon of different size around the central point or central place until the greatest number of hexagons overlapped each other. These central places became the sites where the aggregate distance between all settlements is reduced and the greatest number of goods can be provided at one location.

This led to the minimising of movement between locations and was seen as efficient in terms of the economic landscape. Losch identified that this 'economic landscape' would lead to rich and poor sectors existing; six would be 'city rich' and contain the central place for the purchase of many products and thus satisfy demand, while six 'city poor' areas would also exist that were lacking in goods, services and have a lower population density because they were so underserved.

Losch's work looks at supply and demand, and tries to marry the two together on the economic landscape and is a useful theoretical starting point complementing Walter Christaller's central place theory. Both were concerned with point, line and area as the foundation of spatial analysis.

c) David Smith's space cost curve model

David Smith (1971) investigated the spatial relationship between costs and revenue. Like Weber, he used isodapanes to identify the optimum location for production and assumed that profits would decrease from this location.

Smith developed the work of Rawstron and constructed the space cost curve to represent a cross-section through an iasdapane map.

In the first model (see Figure 19a) revenue decreases with distance from the central point of maximum revenue and is represented by a downward sloping line. The cost of production is constant and transport costs increase with distance from the central point. Profit decreases from the central point of maximum revenue and the limit or margin of profitability is reached where the space cost curve falls below the horizontal line representing costs of production.

If, however, the point of least-cost location is considered and revenue is constant, the cost of production increases from that location until the spatial limit of profitability is once again met (see Figure 19b).

Where both costs are variable, the size of the market that could be served profitably alters once again as the spatial margin moves its location (see Figure 19c).

Such spatial margins will vary from industry to industry, but the concept is a useful one in showing how an entrepreneur may look at

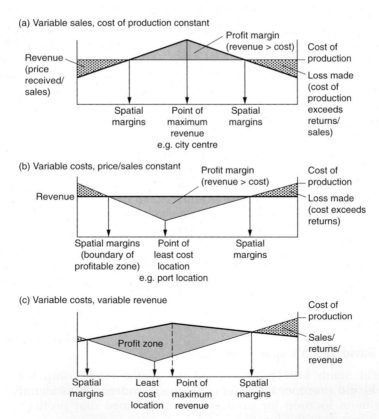

Figure 19 Space cost curve model. *Source*: Nagle (2000).

the marketplace. Identification on the spatial limit of profitability is important and allows the skilled entrepreneur to look at ways to increase market share and profitability at the same time. The effects of external influences, such as government, can also be incorporated into such models and may lower costs of production with the introduction of such incentives as tax breaks or the reduction in rates on premises bringing down the costs of production. Such incentives can be seen as a way of keeping a company, or encouraging new companies to locate in a particular area as the difference between costs of production, revenue and spatial margin of profitability increase.

The model Smith developed, depicting the influences on the location of industry, is significant in that it goes beyond 'economic man' and identifies the limits within which the entrepreneur works and moves beyond the point of profit maximisation.

d) Alan Pred's behavioural matrix

Pred's work is different in that it recognises that not all entrepreneurs are optimisers concerned with profit maximisation. The term satisficer can be used to describe those who are satisfied with a level of profit that is less than the maximum. Psychic income, possibly in the form of an attractive environment, proximity to a golf course or even a school with a good academic reputation, are factors of worth to such an entrepreneur. All enhance the standard of living and the quality of life, and may be considered worthwhile as an alternative to profit maximisation that may mean location in an inner city or a dirty and polluted environment.

The behavioural matrix devised by Pred (see Figure 20) has two axes, with one indicating the quality and quantity of information available to the entrepreneur and the second indicating the ability to use.

The location chosen on the matrix indicates the amount of knowledge assimilated, with a left-biased location indicating poor choice and a right-biased location indicating a more appropriate and knowledgeable choice. The map accompanying the matrix also indicates the spatial margins of profitability, the optimum locations and the sites chosen. Not all factories will succeed and two are likely to close as they fall outside the parameters of profitability.

Position on the matrix, therefore, indicates the behaviour of the industrialist or entrepreneur and can help explain deviations from the optimum location as well as indicate the likely level of success of an enterprise.

e) Kirk's conceptual model of the behavioural environment

In the 1960s William Kirk borrowed ideas from Gestalt psychology. They are different to the theories postulated by Pred, in that the

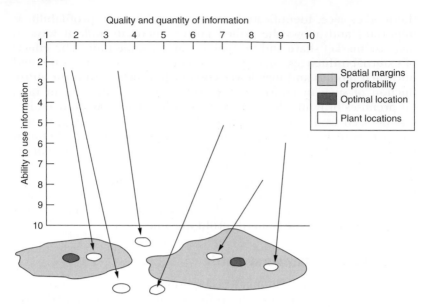

Figure 20 Pred's behavioural matrix. *Source*: Guinness and Nagle (2002).

foundation of his concept lies in the view that the objects we view in the world have a different meaning when viewed as a whole rather than as individual items. Consequently, the geographic environment in which people work can be classed as a unified study. However, within this environment there are two subdivisions: the phenomenal environment, this being the real world, and the behavioural one, being how it is perceived in the mind of the individual (see Figure 21).

As people are in contact with the physical world and therefore the phenomenal environment, they have the power to change it and to be changed by it. Behavioural reactions will depend on motives, preferences and modes of thinking associated with the cultural background from which any person comes. Different meanings will exist in different cultures and as new ideas develop the view of the opportunities within an environment may change.

In industrial societies where new ideas, innovations and opportunities abound, the prospects of the behavioural environment are seen by Kirk as ways for the entrepreneur to exploit both the physical and human environment to his advantage and lead to new economic initiatives.

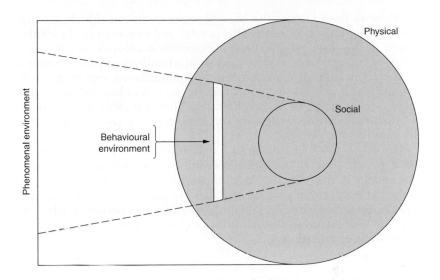

Figure 21 Kirk's conceptual model of the behavioural environment.
Source: Pryce (1977).

4 Do These Theories Take Account of the 'New Industrial Revolution'?

None of the above theories have been postulated since the shift in production from the old to the new centres of industrial production.

However, all of these theories can be seen to have a value in that they provide ways of looking at the industrial landscape. The work of Hotelling, for instance, recognises the importance of competition within the marketplace and that entrepreneurs can co-exist and operate in a profitable manner. This can be illustrated by the growth in number of overseas car manufacturers operating in the UK which are still profitable. Their location also needs further explanation in that the UK is seen as a staging post for the EU market, receiving volume imports from the NICs and thus protecting the company's share of the European market.

David Smith's work is more directly concerned with industrial location and looks at costs of production and the importance of transport costs. This work can be linked directly to the growth in industrialisation of Japan and the Far East as the significantly lower costs of such economies can be seen to outweigh the additional distance involved in serving the economies of the Western world. In contrast,

Pred's work does look at the entrepreneur as an individual and not just as an optimiser of profit. This is valid and can assist in the explanation of the location of industry, especially small-scale production, when an entrepreneur may be operating a family enterprise serving a local market. Even though there may be potential for expansion, it is not taken as it may involve relocation or expansion beyond which the entrepreneur may not be comfortable. This also takes into account the work of Kirk who recognises the importance of psychology in the decision-making process.

The importance of industrial location theory lies in its ability to make people look at the landscape and to interpret what they see. The change in the industrial landscape with the growth of the nations of the Far East can, in part, be explained in theoretical terms, but the uniqueness of each individual company and the economy in which it operates may see other historical factors being of similar significance in the location of industry.

5 The Global Shift in Manufacture

Industrialisation has spread from Europe and the USA and now incorporates, in particular, the nations of the Far East and to a lesser extent South America. Africa remains underdeveloped. Industrialisation in the Far East saw Japan being the first of these nations to develop, and subsequent growth in Taiwan, South Korea and, more recently, China has seen the region become the major centre for economic growth, hence the term the 'tiger economies'. In South America, Brazil has industrialised but the spread of economic activity and wealth has not been as significant as in the Far East.

CASE STUDY: JAPAN

Japan was the first nation of the Far East to industrialise in the late nineteenth and early twetieth centuries and, although lacking in the necessary raw materials, developed a highly sophisticated car and shipbuilding industry. Japan's defeat in World War II left the country both poor and weak. The co-operation between government and industry helped the country to overcome these obstacles. With investment in modern technology, the re-investment of profits and a low-wage economy, Japan was able to develop an industrial foundation that enabled it to turn defeat into economic prosperity. The Japanese workforce worked a lean and efficient industrial programme with co-operation at all levels within industry, and this was evident in companies such as Toyota where employees were guaranteed jobs for life and in return were expected to show loyalty to the company. This produced very competitive goods in the world's marketplace and enabled Japan to prosper. This prosperity also fuelled a greater home demand among the country's 120 million inhabitants for the products being made and a cycle of economic growth took place.

The car and the shipbuilding industries were taking away trade from traditional sources of these products, e.g. the UK, as they were produced more competitively and economically. Similar growth was seen in the electronics industries. All products could be easily exported by sea without major cost implications that would have made their products unattractive to the world market.

By the 1970s Japan had replaced the USA as the world's leading car manufacturer. It became a maker of low-cost bottom-of-the-range cars, able to compete on the world market even with the additional transportation costs involved in sending the vehicles to the mass markets of Europe and the USA. Efficient production techniques using high technology on the assembly line, low wages and government assistance were the keys to their success, and companies such as Datsun (now Nissan) and Toyota were able to become world market leaders.

Similar success was seen with the shipbuilding industry. Postwar shipbuilding was dominated by Europe and the USA, but today only two of the top 22 builders of bulk carriers, tankers and containerships over 100,000 tonnes are located in either area – these being General Dynamics/NASSCO and Northrop Grumman/Avondale in the USA. Of the remaining 20, 10 are Japanese including Mitsubishi, five are Korean including Daewoo, four are Chinese and one is Taiwanese. In contrast, the River Clyde in Scotland has seen a decline since 1946 in shipyards from 42 to three in 2001 and a fall in tonnage of ships produced from 475,000 tonnes in 1955 to zero in 2001, although new orders have since been received.

Japanese success led to a crisis in the car and shipbuilding industries of both Europe and the USA, especially when the range and quality of products made in Japan and other Far Eastern nations increased and put further pressure on already ailing industries. The threat of import quotas for cars into Europe led to a strategic policy shift by Japanese car manufacturers who saw that the only way to secure access to the lucrative markets was to produce cars in Europe themselves. The UK, in particular, benefited from this, and Nissan, Toyota and Honda all now have production lines assembling over 500,000 cars a year, mainly for export. Similar sanctions could not be made for ships which tend to be built to an individual design and are built to order.

6 High-tech and Footloose Industries

High-tech industry has access to the most modern methods of production and is often associated with micro-electronics. It covers a wide range of products and industries, from the production of space vehicles and medical instruments to biotechnology and pharmaceutical products. Such industries are distinguished by their complexity and their reliance on research and development (R&D). It is typical for such industries to spend 1% of their income and employ 5% of their workforce on R&D.

Footloose industries are not tied to a particular location and are therefore mobile; many high-tech industries are able to operate in this way as *knowledge* is their raw material. Consequently, the success of any such enterprise is dependent on the ability to recruit and retain highly skilled research scientists, technicians and engineers. To do so, an attractive environment, preferably in close proximity to a university with laboratory and research facilities where students working in the applied sciences provide the next generation of employees, is an ideal location.

The work of Gripaios highlights these points in research undertaken in 1989 whereby companies interviewed ranked five locational factors according to their company's requirements. The aggregate scores are indicated in Figure 22.

Areas of high-tech industry have grown close to universities such as Cambridge in England and in the sun-belts of the Mediterranean and the Pacific coastline of California, USA. Centres such as El Valles in Barcelona, Spain and Silicon Valley near San José, California, USA are typical centres of high-tech growth.

In El Valles high-tech labour is concentrated adjacent to a Science and Technology Incubator Park (Parc Tecnologic del Valles) and the nearby university specialises in educating students in the skills of applied science. The Catalan government has also invested heavily in support infrastructure with motorway development and improvements at the international airport being built. Such locations also encourage the clustering and agglomeration of similar companies and associated supplier and service providers. The planned Southern

Locational factor	Ranking
Pleasant working environment	63
Access to good workforce	51
Available sites	50
Proximity to other high-tech companies	17
Supportive local authorities	17

Figure 22 Locational factors in high-tech industry.
Source: Gripaios (1989).

European Synchrotron (Accelerator) Radiation facility in El Valles is likely to ensure that this high-tech complex gains further momentum.

Silicon Valley is the location for over 180 high-tech companies employing over 225,000 people. Software, systems, defence and communications are four areas of specialisation in an environment possessing similar climatic conditions to Barcelona and offering a good quality of life. The San José State University provides research opportunities for more than 30,000 undergraduates and postgraduates, and is an essential source for recruitment into high-tech industry.

Such industries have become increasingly significant to the USA, western Europe and Japan as other NIC nations develop an industrial base. In 1995 they accounted for 14% of the UK's and 15% of the USA's and Japan's manufactured output, while in China it accounted for 12.5%; a figure higher than that for either France or Germany. The USA and Japan dominate world trade in high-tech industrial products – with Japan overtaking the USA in 1991 as the world's leading high-tech producer. This situation has since been reversed due to deflation within the Japanese economy and accounted for less than 23% of world production by 1995. Meanwhile China's share of the market had risen to 6% by the same date.

Industrial fortunes can be seen to evolve over time. This emphasises the dynamic nature of industry and the significance of techno logical development within an industrial cycle.

The effects of this can be best seen by studying how such a change in industrial location can affect particular cities that have evolved in different ways.

7 The Industrial and Post-industrial City

The term 'post-industrial society' and hence city, is associated with the research into Western society by Daniel Bell who, in 1973, emphasised the significance of technology, rather than the mode of production, as a major force in shaping society. It notes the transition between the early capitalist forces of industrialisation of the nineteenth century to the post-industrialist and late capitalist society of the late twentieth century, where unskilled labour in the production process is reduced and replaced by the increasing use of technology, not only in the factory but also in service industries and general administration. Technological advances dominate factory production at the expense of labour, and R&D institutions gain an increasing significance. This leads to a fundamental change in the character of the workforce with an increase in white-collar workers and a decrease in blue-collar workers who would normally have worked on the factory floor. This reduction in the number of people directly concerned with the manufacturing process and the rise in service employment provision directly affects the structure and fabric of the city.

CASE STUDY: A COMPARISON BETWEEN BIRMINGHAM AND VANCOUVER

Each city's history is unique. Some have evolved through the exploitation of raw materials and the subsequent development of industry, while others have come about from a very different starting point.

Birmingham's history dates back over 600 years while that of Vancouver in British Colombia, Canada only dates back to 1886.

Vancouver

Vancouver's origins coincide with the decision to choose the Burrard Inlet as the terminus on the western coast of Canada for the trans-Canada Canadian Pacific Railway.

Vancouver's growth was not based on the exploitation of minerals and the manufacture of secondary product. The significance of timber and its associated industries cannot be underestimated, but it has developed in a fundamentally different way to Birmingham. Many industries are associated with Vancouver's port, with engineering, ship building, oil refining, aircraft manufacture, fishing and canning all being industries located in the city. However, Vancouver also houses commercial offices and retailing, which date from the 1960s and, as a result, the city can be classed as a post-industrial city. It has been stated that in the 1960s 'a new ideology of urban development was in the making' in the city of Vancouver.

The growth in quaternary or service sector employment in Vancouver has seen the city become a major centre for both national and international banking, accounting and law, together with software development and biotechnology. The central area has 60% of the region's office space, with many of the headquarters of the mining and forest industries associated with the exploitation of the province of British Colombia's natural resources being located there. Vancouver's spectacular setting – it is surrounded by water on three sides and the Coast Range Mountains rise 1500 metres behind the city – has helped it become very much a 'West Coast' city, on a par with San Francisco and Los Angeles. In all, the quality of life on offer in the city is seen to be of paramount importance and, even though the city has doubled in population to 1.6 million in the 30 years to 1991 and has a projected increase to 3 million by 2021, this emphasis on quality remains as significant as the generation of new economic activity to support further growth and development.

Policies associated with the liveable city ideology stem from shifts in the political, economic and socio-cultural ideals dominant in the late 1960s and remain mainstream today.

Vancouver has become very much a Pacific-rim biased community with large Chinese, Japanese, Korean and Vietnamese communities. The Chinese immigrants came in the 1800s for the Gold Rush and to build the railways, while the Japanese arrived to work in logging and fishing industries. The Korean and Vietnamese communities reflect the influx of refugees from the wars of the 1950s and 1970s, while immigrants from Hong Kong reflect the handover of that country to Chinese rule. Today, the city's population is approximately 60% Asian and links with other Pacific nations are important to the city's continuing prosperity.

Birmingham

Birmingham's pattern of growth is somewhat similar, but started some 40 years before the inception of Vancouver. It was not until 1905 that growth accelerated with 100,000 people migrating to the city in the following 7 years. In Birmingham, terraced housing spread to the north, east and south into the former self-contained villages independent of the city. This was added to by massive post-World War I growth when the city further encroached to the north, east and south absorbing many more semi-rural centres such as King's Heath, itself formerly part of the county of Worcestershire.

Birmingham's growth was based on metal and the skills of its workforce in the numerous metal trades. Such was the importance of manufacture that 53% of all UK patents registered since World War II were generated in the city. In Birmingham, the car industry has, in the past, accounted both directly and indirectly for up to 70% of the manufacturing base.

Birmingham of the 1960s was in a period of urban renewal, which was criticised for putting the car before the needs and requirements of people. Birmingham city centre was dominated by the car and people were expected to use underpasses to cross major roadways rather than the reverse. The central area gained a very negative reputation and with the decline in manufacturing output, in particular of the car industry, there was concern that the city could go into terminal decline.

Hobday Ltd is an example of how companies have had to change to the new post-industrial climate.

Hobday Ltd, situated in Aston, Birmingham, is typical of many small metallurgical companies based in the city, being a company specialising in the manufacture of shop display fittings in metal, plastic and wood. The workforce of 30 was semi-skilled, employing welders, chromium platers and carpenters. Since 1991 the company ceased manufacture and has become an importer of shop fittings. Lower costs in the Far East, in particular China, made such production uncompetitive.

- Sixty-five per cent of the company's product now originates in China with the EU accounting for another 25% of its imports.
- The cost of such imports is less than the cost of the raw materials Hobday's would use for manufacture.
- Today only 'speciality' goods, on which a profit can be made, are manufactured.
- The workforce now has a lower skills base and is largely concerned with warehouse and distribution compared to manufacture.
- If the company had not decided to import its product and had attempted to continue to manufacture, it is likely that the company would have folded and the 20–25 employees would have lost their jobs.

Sales today are via a mail-order catalogue and website that has enabled the company to survive.

Now, like Vancouver, Birmingham has become a centre of service and administration employment. It is no longer reliant on manufacturing, although it is still the home of car makers MG Rover and Jaguar and component manufacturers such as Lucas, GKN and chocolate maker Cadbury. New companies such as Marconi, Hitachi and Siemens are also resident in the city. Birmingham now has the largest professional and financial sector outside of London, and the West Midlands has been voted the best region for relocation among members of the Confederation of British Industry (CBI).

- Gap Inc., the US clothing company, has built two warehouses and an office complex on an 80-acre site in the city.
- In 2001, 92 inward investment projects created 5500 new jobs.
- German investment in the West Midlands has reached £884 million and created 51,000 new jobs.
- The USA is the chief inward investor.
- City law firms now contribute £800 million to the local economy, roughly 40% of which is generated outside the region.
- Instead of losing jobs at a rate equivalent to more manufacturing jobs lost between 1980 and 1983 than Scotland and Wales combined, Birmingham is now reversing that trend.
- Industrial closures still continue: Alstrom, the Anglo-French engineering company, is reducing production of trains at its Washwood Heath plant, with 1100 jobs being lost.
- Smaller engineering and component companies suffer from overseas competition where low-wage economies exist. The relative strength of the pound also affects such companies.

Like Vancouver, Birmingham has a large migrant population. Those in Birmingham stem largely from the migrations from the

West Indies in the 1950s and later the peoples of the Indian sub-continent to fill vacancies in industry. All have Commonwealth links with Britain. There is also a Chinese quarter and a large Irish population.

Birmingham's renaissance continues. September 2003 saw the opening of the new Bull Ring shopping complex replacing what had become a symbol of the city's physical decline. It is Europe's largest enclosed city centre retail complex covering a 26-acre site. The £550 million development completes 20 years of urban regeneration and will house 50 new shops, which are expected to attract 30 million shoppers per year from a catchment of 7.2 million people who live within a 1-hour drive of the city. Together with the Symphony Hall, the National Indoor Arena and the National Exhibition Centre located adjacent to the international airport, which itself is seeing an annual growth rate of 12%, Birmingham is now on a par with Vancouver as a post-industrial city. Their evolutions have been different but today their similarities outweigh their differences which would have been the case even 20 years ago.

Questions

1. How does the work of Smith, Pred, Kirk and Hotelling help in the understanding and explanation of industrial location?
2. Describe the effect on the UK that has occurred due to the shift in manufacturing to the other parts of the world, particularly NICs.
3. How have cities such as Birmingham adapted to the new location of industry?
4. How are the location factors for modern industries, such as those in the high-tech sector, different from those for many traditional industries.
5. How have the cities such as Birmingham adapted to the decline in manufacturing?

4 Tertiary and Quaternary Activity

The tertiary sector of industry is concerned with the provision of a service. This can be in the form of retailing or the provision of services to the people within the community whether at local or national level. Quaternary industry, a subdivision of the tertiary sector is more concerned with financial and insurance activities, research and development, media and government agencies.

1 Tertiary Activity

a) Retailing

The sale of goods produced by the secondary sector of industry and the provision of services is a major source of employment, especially in the MEDC nations where greater prosperity and disposable income allow people to purchase products non-essential to life.

Convenience, comparison and luxury goods

The products of industry can be classified as either convenience, comparison or luxury goods. Convenience goods are foodstuffs and other items purchased on a daily basis such as newspapers, which are usually purchased locally, while comparison goods are more expensive and people tend to compare items before deciding on a purchase, e.g. shoes or clothing. Luxury goods require a greater financial outlay and tend to be less frequent purchases, e.g. cars. Each product has a certain market area that can be identified by its threshold and range. Range is the distance people are prepared to travel to make such purchases or the distance the supplier is prepared to travel to deliver, and the threshold is the minimum number of customers needed for an enterprise to be successful.

To service these different categories of goods, a retail hierarchy or order has developed.

- Convenience or low-order goods such as bread and milk are provided at a local level as they have a low threshold and range. These would be first-order settlements and would often be of the corner shop or local village shop.
- Where shops are interspersed within an area or where groups of shops congregate, often on converted ground-floor premises, a second order of retailing would exist. These would be neighbourhood units and contain small supermarkets and a dominance of privately owned shops. All goods sold in the first-order shops would be available in the second-order retail facility.
- Such retail facilities serve a local area and are greater in number than the larger third-order suburban centres. These are larger in size, sell a greater number of products of lower (convenience) and middle order (comparison), and serve a larger threshold population who travel a greater distance, or range, to visit the centre. These centres may vary in size and their retail facility may reflect the requirements of the local population. In the UK they are often linear in shape following the line of the road on which they are located.
- Dominating all of these retail centres are the central business districts (CBDs). These are dominated by comparison, specialist and luxury goods shops. These are the fourth-order settlements and are fewer in number, but more economically important as they serve an even larger threshold population, many of whom travel a greater distance to reach this retail facility.
- The 1980s and beyond were dominated by the rise of purpose-built out-of-town shopping centres serving not just a town or city but a whole region. Arguably, these may be classed as fifth-order facilities as they provide a further tier of retail facility in the urban environment.

The work of the German geographer Walter Christaller recognised that such a pattern exists in a landscape. His central place theory, which looked at the distribution of settlements over the landscape, led to the classifying of settlements according to their size and the range of services offered in each. Starting with the smallest settlement and retail facility, through to the largest settlement with a concentration of retail facilities and services in the CBD, all settlements and retail centres could be classified.

His theory used a series of assumptions that stated that:

- A flat isotropic plane existed with ease of access in all directions.
- Transport costs were proportional to distance.
- Population was evenly distributed over the isotropic plane.
- Resources were evenly distributed over the isotropic plane.
- To minimise distance travelled, goods and services were always obtained from the nearest central place.
- All customers had the same purchasing power and made similar demands for goods to be purchased.

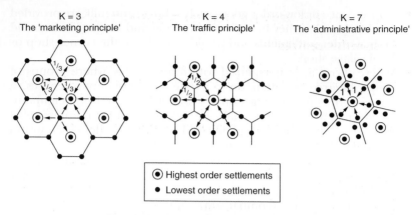

Figure 23 Christaller's three patterns of trading area. *Source*: Nagle (2000).

Christaller considered that the circle was the ideal shape for the sphere of influence of a central place. However circles do not fit together when their circumferences touch and there are areas outside the boundary of the circle. According to Christaller these would be unserved areas. If the circles are redrawn leaving no areas excluded there would be overlap between competing central places. To solve this problem Christaller modified the circles and created a pattern of hexagons with no unserved or overlapping areas. This shape proved ideal for superimposing the market areas for different orders within the hierarchy of retail settlements.

Christaller assumed that people who lived in a hamlet would obtain low-order goods from that hamlet and travel to nearby villages or towns to purchase those goods or services which required a higher threshold and were not available locally. Therefore, the order of the retail centre could be identified by the type and number of functions it possesses, its market area and its population size.

Christaller produced three patterns of trading area called the $k = 3$, $k = 4$ and $k = 7$, where k is the number of places which are dependent on the next highest order central place and where the whole population is served.

Out-of-town shopping centres, popular since the 1980s, were not part of the human landscape at the time of Christaller's work. They are an addition brought about by technological innovations and social developments such as the home freezer and private transport enabling people to buy in bulk and store and to travel beyond their local centre for their purchases.

Christaller's work can be set against reality by looking at examples of retail centres to see if a hierarchy actually exists.

CASE STUDY: RETAILING IN BIRMINGHAM

First-order centres

In Birmingham, as in all large urban settlements, this pattern can be recognised. There exists a pattern of first-order corner shops throughout the city, especially where the older Victorian terraced buildings still exist, although their numbers are decreasing due to redevelopment, changing shopping patterns and increased mobility.

Figure 24 Settlements in Birmingham.

Second- and third-order centres

Local shopping parades such as the one in Haunch Lane, King's Heath, Birmingham are typical of second-order centres. It contains 15 shops, all privately owned and selling convenience goods. The third-order retail centre of King's Heath itself focuses on a linear facility following the line of the A34 Alcester Road and its

Figure 25 King's Heath: 'low-status' shopping centre.

side roads. King's Heath is well served by local bus routes but lacks car parking space. It is within easy walking distance for many who benefit from the shops owned by both private individuals and national retailers, selling all the products found in the lower order centres and higher comparison goods. The shops are contained on the ground floor of converted Victorian and later buildings, and in purpose-built structures such as those occupied by retailers such as WH Smith, Safeways and Sainsbury's. Lack of space for expansion has led to Sainsbury's building a larger purpose-built supermarket 2 miles south of King's Heath at The Maypole on the edge of the city. At this location, a larger store with car parking has been built at a junction that enables ease of access from a greater area than possible in King's Heath.

King's Heath is typical of many third-order settlements throughout the UK providing easy access for local communities to a medium range of retail facilities. They reflect the socio-economic composition of the area that they serve and may possess retail facilities reflecting the community's greater need. Of the 8418 wards, or local government electoral areas in England, King's Heath is ranked 1197, with 1 being the lowest and most deprived. In King's Heath there are in excess of 230 shops, 30 of which are categorised as chain stores. These include Boots, Burtons, Dorothy Perkins and Woolworths. The presence of 10 charity shops and those whose prime function is cashing other people's cheques is indicative of the lower levels of income earned my many residents of King's Heath. Similar outlets may not be available in higher status retail centres of Birmingham such as Shirley or Acock's Green reflecting their higher status catchment area.

Sparkhill, a similar retail area to King's Heath, also reflects its cultural diversity with many Asian shops selling food and clothing, and meeting the needs of its ethnic community.

Fourth-order centres

The central business district

Traditionally it was the CBDs that were the major retailing centres for towns and cities. Birmingham's CBD is no exception and can be considered to be the major retailing centre not only for Birmingham but also the West Midlands conurbation. It possesses department stores and outlets for all of the major UK retailing companies.

Figure 26 Inside Birmingham's new Bull Ring Centre.

Figure 27 The Mailbox: 'high-status' shops and offices in Birmingham.

Following a period of stagnation and the loss of all but one major department store, the addition of the new Bull Ring Centre on a 26-acre site, at a cost of £550 million, will ensure that Birmingham remains in the top league of retail facilities in the country. It is estimated that within a 60-minute radius of the city there are 7 million consumers with an annual spending budget of £4.1 billion. The addition of 150 shops in the new Bull Ring Centre together with the previous shopping facility is expected to attract up to 30 million shoppers a year. Debenhams and Selfridges have opened new department stores, and new retailers such as Zara, Nike and Lego have opened stores in the city for the first time. The new centre will generate £40 million a year in rents for the owners of the development, Hammerson, Land Securities and Henderson Global Investors who together formed the Birmingham Alliance to enable its redevelopment. It has also created 8000 jobs and now boasts as having the best concentration of shops outside of London's Oxford Street housed in the largest undercover retail facility in Europe.

The CBD also benefits from the redevelopment of the former Royal Mail sorting centre that has now been converted into The Mailbox and houses high-status retailers such as Harvey Nichols and Armani. It is located west of the Bull Ring on the opposite side of New Street rail station and has brought about the extension of the retail facility in the city. It also borders Gas Street Basin and Broad Street, the city's major recreational, sporting, concert and conference area.

The re-invention of the CBD

The re-invention of the CBD recognised the impact that out-of-town shopping centres were having on the traditional role of the CBD as the major shopping area of a city. Many chain shops and department stores had either relocated or kept smaller shops in the old city centre and specialist or independent retailers who often relied on the presence of such companies went out of business, leaving abandoned shops, which often went into decay. The built environment thus also went into a state of decay.

The redevelopment of cities such as Birmingham and Manchester after the terrorist bombing of the Arndale Centre are enabling such CBDs to be restored to their former status. This rejuvenation was aided, in part, by concerns from government on the affects of building out-of-town centres not just on the CBD but also on the amount of greenfield land that was being taken for such purposes when brownbelt land within city areas was available for development.

Fifth-order centres

Out-of-town shopping centres

The trend from the 1970s onwards was to develop purpose-built, out-of-town retail centres on greenfield sites at locations with ease of access from its market area by private and public transport. This was a contributory factor in the decline of the retail facility in the CBD of Birmingham and many other city centres. Inner-city congestion with poor access and parking also accentuated this trend and made such alternatives desirable, as did lower land values compared to those in the CBD where rates and rents were also higher. This allowed individual shops to use large areas of floor space, and to stock a wide range and large volume of goods for their customers and have possible room for expansion.

Since 1980, 80% of all new shopping space has been on out-of-town sites. Metrocentre in Gateshead, north-east England was the first regional shopping centre to be developed and others have included Meadowhall in Sheffield and Bluewater in Dartford.

Merry Hill out-of-town shopping centre

Merry Hill shopping centre in Staffordshire dates back to 1985 when the first of five phases opened. By 1989, a retail facility covering 1.5 million square feet was open to the public. It is located between the Dudley and Cradley Heath in the West Midlands on land surrounded by the A491, A461 and A4101 roads with access to the West Midlands motorways. Merry Hill boasts that it can be accessed from Birmingham and other settlements within the region at peak times in 30–45 minutes. It houses over 200 shops with all the major retail stores are represented at Merry Hill in a purpose-built undercover structure and it also offers 10,000 free car parking spaces for its customers. A measure of its success is the 21 million annual visits every year made by the public, 70% of which are made from a within a 10-mile radius of the centre, with a high percentage making weekly visits.

In 2002, the developer Chelsfield plc, owner of Merry Hill, bid to have the shopping centre, or mall, rezoned as part of a new emerging town centre linked to the nearby Brierley Hill high street. This followed a decision in the late 1990s that proposals for an extension to the centre should be turned down because it did not occupy a town centre location. The bid was rejected by the West Midlands regional planning guidance group as linkages were deemed to be insufficient for such a rezoning to be approved. It also rejected any further retail development until the area's infrastructure was improved.

Although MEDC nations have a greater disposable income, the retail hierarchy can also be recognised in LEDC nations.

CASE STUDY: ZIMBABWE RETAIL HIERARCHY

Like the UK, where London is the highest order shopping and retail facility, in Zimbabwe the same can be said for the capital Harare. Here the highest order goods are available in purpose-built centres both in the city and on the fringe of the city such as at Bluff Hill. In hierarchical terms, Bulawayo, the second city, possesses many of the same retail outlets but beyond these two cities the retail facility is much poorer.

Only 32% of the population is urban and the average income for Zimbabwe is below $200 a year. In the rural areas, access to retail facilities is limited and will normally mean a long walk and a bus journey to the nearest settlement where trading is largely dominated by open-air markets. Farmers either sell their produce at the side of the road or at such markets to gain money to purchase goods that they are not able to provide for themselves.

To ease the problem of distance, growth points have been established in many of the rural areas. In the Hondi Valley in eastern Zimbabwe, near to the Mozambique border, such a facility has been built midway along the length of the valley. It contains 10 retail outlets selling farm produce and basic domestic needs such as foodstuffs, material, clothing and soap. There is also a post office, a petrol station and an airstrip, and it is served by two daily buses from the nearby town of Mutare. Although basic by Western standards, it does give people better access to retail products than before. The recognition of the lack of adequate retail facility in the city also led Mobil to establish retail outlets in all their southern Africa petrol stations, selling both tinned and fresh produce similar to those facilities now found in many UK petrol stations. This has also increased many people's access to basic food products that had otherwise been denied them.

The street market, however, still dominates rural retail life. Stalls selling shoes made from rubber from discarded tyres, baskets made from woven straw and locally grown produce are often the only major source of income for the Zimbabwean farmer. Co-operatives using the local talent in wood and stone carving, in pottery and cloth are also sources of income to compliment the income of the poor.

The retail facilities of countries such as Zimbabwe do possess a hierarchy or order to them, but it differs significantly from its counterpart in the more wealthy nations. With 70% of the population living below the poverty line the demand for a more complex pattern of retail facilities is missing. The level of disposable income in the country is clearly reflected in the retail facility found in Zimbabwe, which is dominated by the needs for convenience goods, in particular food and clothing. Beyond these necessities, demand is limited and is reflected in the smaller range of goods available to the people of Zimbabwe.

Figure 28 Zimbabwean street market selling fresh produce as cash crops.

b) Transport and the aviation industry

The effects of growth in the tertiary sectors have meant a huge increase in air travel and employment in associated aviation industries:

- Over 1.6 billion passengers travel by air annually for business or leisure purposes and this figure is expected to be in excess of 2.3 billion by 2010 and double by 2020.
- It generates 28 million direct, indirect or induced jobs worldwide and in Europe on average 4000 jobs per million passengers carried are created and sustained.
- Air transport in Europe comprises of more than 130 airlines serving over 450 airports and generates receipts of €700 million per day from the tourist industry.
- The location of an airport is important in enabling an area to develop as the economic ripple effect attracts other industries and business activities to an area.

London's Heathrow Airport

This is particularly evident at London's three major airports who handled in excess of 109 million passengers in 2002, of which 63.34 million used London's Heathrow Airport. The tertiary industries allied to the catering for this number of passengers generated over £5 billion in wages for the British economy. It supports 250,000 jobs in the UK, of which 108,000 are found in the area of west London near to the airport and 68,000 are actually based at the airport itself.

Heathrow is the third busiest airport in the world and the largest international airport in the world transporting passengers from one country to another.

Figure 29 London's Heathrow Airport.

1	Atlanta	76,876,128
2	Chicago	66,565,952
3	London Heathrow	63,338,641
4	Tokyo	61,079,478
5	Los Angeles	56,223,843
6	Dallas	52,828,573
7	Frankfurt	48,450,357
8	Paris Charles de Gaulle	48,350,172
9	Amsterdam	40,736,009
10	Denver	35,651,098

Figure 30 The world's busiest airports, 2002. (London Gatwick is 22nd and London Stansted is 65th.)

Heathrow is also Britain's largest cargo port by value of goods imported and exported, carrying in excess of 1.25 million tonnes of freight annually. Together with the six other airports operated by BAA plc, approximately 30% of UK exports worth £50 billion were transported by air in 2002/3.

To maintain its position as the premier centre of aviation in the world, proposals have been put forward to build additional runway capacity in the London area. British airlines, such as British Airways, Virgin Atlantic and British Midland, want to see an additional runway constructed at Heathrow as it serves 90 airlines and over 160 world-wide destinations, but alternative proposals have also been put forward for new runways to be built at Stansted and Gatwick instead. Andrew Cahn, Director of Government and Industry Affairs at British Airways said that 'if Heathrow is not chosen, then the government is condemning UK aviation to go the way of the shipbuilding and motorcycle industries', as an alternative location would not meet the needs of the industry and airlines would be likely to relocate to other European airports.

Gatwick and Stansted are not seen to be as suitable locations as they tend to serve the UK holiday and low-cost carriers rather than scheduled international airlines. It is predicted that the number of passengers wishing to fly from south-east England will increase from the current 110 million to 300 million plus passengers by 2020 and that Heathrow is the favoured airport for both passengers and airlines alike.

Opposition to Heathrow's expansion is based on congestion, pollution and the demolition of people's homes to build a new runway. An additional runway would require construction on greenbelt land and the possible demolition of the villages of Harmondsworth and Sipson. The exposure of people living under the flight path to nitrogen dioxide (NO_2) would be increased by 35,000 if a new runway was constructed at Heathrow compared to only 35 at Stansted, as the new runway would add an additional 500 flights a day.

The need for further capacity in south-east England is not disputed and its significance for employment and earnings cannot be underestimated. The government decided that the location of the new runway for south-east England would be at Stansted, but that Heathrow may be allowed to build a third runway between 2015 and 2020 provided that aircraft emissions are reduced sufficiently by that time. The possibility of the third runway at Heathrow was welcomed by British Airways especially as in the meantime consideration is being given for the dual use of the two parallel runways which would mean that both could be used for simultaneous take-offs and landings instead of one runway being exclusively used for each. This would enable an additional 60 flights per day to be accommodated catering for up to 20 million more passengers per annum.

The Confederation of British Industry (CBI) believes that the government have made a 'balanced approach' in the Aviation White Paper and that business was 'hugely relieved' by the decision to allow expansion at Heathrow. The Trades Union Congress (TUC) was disappointed that Heathrow was not the first choice for expansion as it would be a greater boost for the economy while opponents such as 'Clearskies', which represents people living near Heathrow, feel that an opportunity has been missed to rule out further expansion at the airport.

Airport expansion is always going to be a contentious issue and the debate about it will continue.

Often associated with airports and universities are footloose industries that need ease of access to transport links and the skills found at such academic institutions. This has led to the rise in business parks in close proximity to such facilities.

c) Leisure and cultural activity as a service industry

Leisure, recreation, culture and the arts are all encompassed within tertiary activity. Their growing significance, as areas of employment, to meet the needs of greater leisure time among the population is being recognised. Similarly, the cultural heritage not only of a nation, but of cities within the nation, is also being recognised. Within the EU this is clearly evident with the nomination each year of a city of culture.

CASE STUDY: LIVERPOOL, CITY OF CULTURE

In a competition for the European Capital of Culture 2008, Liverpool was successful in a competition with five other British cities. Birmingham, Bristol, Cardiff, Newcastle–Gateshead and Oxford had all submitted bids for the accolade but Liverpool's submission was chosen to host the 2008 event.

The capital of culture award is an EU initiative building on the success of the capitals of culture programme that ran from 1985.

1985	Athens
1986	Florence
1987	Amsterdam
1988	Berlin
1989	Paris
1990	Glasgow
1991	Dublin
1992	Madrid
1993	Antwerp
1994	Lisbon
1995	Luxembourg
1996	Copenhagen
1997	Thessaloniki
1998	Stockholm
1999	Weimar
2001	Rotterdam and Oporto
2002	Bruges and Salamanca
2003	Graz
2004	Genoa and Lille

Figure 31 European cities of culture.

In 2000, nine cities joined under a common heading of 'European Cities of Culture': Avignon, Bergen, Bologna, Brussels, Helsinki, Krakow, Prague, Reykjavik and Santiago de Compostela shared the title. Since the millennium, seven cities have shared the title.

The title has already been allocated to EU nations up to the year 2019, although cities have yet to be decided.

In 2005 the capital of culture is hosted by Ireland, followed by Greece and Luxembourg in the following 2 years. As it is not likely for the UK to host the event again before 2020 at the earliest, Liverpool has gained a significant economic advantage with this development. Inward investment of £2 billion is expected to enhance the city in the run-up to the 12-month event.

Liverpool's location on the western side of the UK with easy access to the North Atlantic and proximity to the industrial heartland of the country secured its growth. Owing to the change in emphasis of trade from Empire to the EU, to which access is less direct from Liverpool, and industrial decline from the 1970s, the city went into decline. Its population fell from over 700,000 in the 1960s to its current figure of approximately 400,000. It became more famous for The Beatles and football rather than its port and associated industries.

However, the city's waterfront is one of the UK's 30 World Heritage sites and is dominated by the Liver Building, the Cunard and the Port of Liverpool building that are symbols of its past prosperity. Its decision to bid for the capital of culture in

1998 was seen as a means by which the city could reinvent itself as a post-industrial city.

The success of its bid means that Liverpool is likely to gain an £200 million in tourist revenue in the build-up to the event and as visitors are likely to move beyond the city and explore the north-west of England creating a further 3000 jobs. A report commissioned by Liverpool City Council by ERM Economics to examine the socio-economic impact of the award shows that:

- The planned developments will reinforce the city's roles as a regional shopping centre, a UK and European tourist destination, and cultural heritage.
- Employment in the cultural sector – tourism, sports, heritage and creative industries – could grow by 14,000.
- Between 2003 and 2008, £2 billion will be invested in the cultural environment of the city.
- The cumulative effect of the award would be 1.7 million more visitors spending over £50 million per annum.

This will build on the recent economic and cultural developments that have taken place including the FACT Centre, the UK's only exhibition and performance space specialising in film and digital art, and the opening of the Liverpool Tate Gallery. New projects for 2008 include:

- Fourth Grace, a new building on Liverpool's waterfront.
- A museum of comedy.
- 'The Cloud', a futuristic design housing the World Discovery Centre at the Central Library.
- A new interactive museum of Liverpool.
- An archive trail showcasing 90 million rare historical records.

All of these are permanent additions to the city and compliment the range of activities that will take place throughout 2008 and draw visitors to the city. To accommodate visitors to such events requires further infrastructural developments in the city. Tourism is currently worth £600 million a year to Liverpool and Merseyside, and is expected to rise to £1 billion by 2005.

By building on its heritage and its 'world in one city' campaign to gain the city of culture bid, Liverpool is moving forward with a new dynamism. Liverpool aims to transform itself with the new investment and urban regeneration that will be forthcoming.

2 Quaternary Activity

a) Financial and insurance services

The impact of new information and communication technologies
(ICT) on the workplace has been significant in both patterns of work
and the location of certain office-based industries. The globalisation
of industry and the impact of ICT has led to a hierarchy of centres
where financial and insurance activities operate, and is dominated by
London, New York and Tokyo from where vast flows of capital radiate
and stimulate the world economy. Whereas these centres were also
the sites of corporate headquarters, this is no longer necessary as ICT
has enabled them to move away from such traditional areas as the City
of London to locations such as Birmingham and Vancouver.

The City of London
The City of London, however, still remains the world's leading
financial centre. It has:

* 300,000 people working in financial, business and insurance
  services
* 72 million square feet of office space
* 481 foreign banks
* 448 foreign companies listed.

The significance of the work undertaken by its workforce highlights the importance of the City of London to the global economy as:

* Thirty per cent of the world's foreign exchange market activity
  takes place in the City.
* It is the world's largest insurance market worth over £14 billion
  (1996).
* London's Stock Exchange is the world's largest trading centre in
  foreign equities.
* It is the second largest fund management centre, after Tokyo,
  managing almost $2 trillion.
* Banks based in the City invest more capital overseas than those in
  any other country, accounting for 18% of global lending.
* $637 billion are traded daily on the foreign exchange market.
* 19% of international bank lending.
* $2555 billion of assets and $2500 billion of metals are traded and
  managed from the City.

The constraints of the overcrowded 'square mile' have led to the construction of a satellite development at Canary Wharf.

Canary Wharf
The development of 86 acres of land with 14.1 million square feet of
office and retail space employing 55,000 people has taken place since

Figure 32 An aerial view of Canary Wharf.

The London Docklands Development Corporation was established by the local Government Planning and Land Act in 1980.

Canary Wharf Tower was finished in 1990 and together with 17 other towers it now houses the headquarters for such banking companies as HSBC, Citigroup and Barclays. The development of the Docklands Light Railway, the Jubilee Line underground extension and London City Airport were significant infrastructural developments that aided the regeneration of Canary Wharf, but ICT has also meant that companies do not have to locate in such close proximity to the City.

Using ICT, work in financial and insurance services is liberated from its former locational restraints and areas outside of London have benefited from the relocation of such services.

Birmingham and the West Midlands now has the largest professional and financial sector outside of London and has been voted the most desirable business location by delegates to a CBI conference. British Telecom, Vodafone, AT&T and the Royal Bank of Scotland have all relocated there.

Similarly, Vancouver benefits from employment in financial and insurance services with many national and international banks, accounting and insurance services locating in the city to serve the mining and forestry industries of the western provinces of Canada. Vancouver's CBD has 60% of the region's office space to accommodate these activities and the headquarters of the mining and forestry companies based in the city.

Vancouver benefits from being the principal city in British Columbia and accounts for 37% of the region's employment. The financial and insurance industries accounts for 10.4% of the work-force accounting for approximately 87,000 people.

Away from the office

ICT also allows people to work in other locations and work from home. This has allowed the locational advantages of locations outside of the city for financial and insurance services to spread nationwide and worldwide for both government and private institutions (see Chapter 6).

Tertiary industry spreads beyond retailing and office-based indus-tries. It also includes tourism, whereby companies and individuals provide services to people. The increase in people's disposable income, increased leisure time and the availability of cheap overseas flights and holidays, together with developments in aviation tech-nology, have made air travel more available to people than ever before.

b) High-tech business parks

Business parks are ideal locations for the high-tech quaternary indus-tries that have developed in the past 30 years. Their dependence on electronics, computing, telecommunications and research institu-tions give them a footloose location. Access to associated industries, research institutions and an attractive working environment become significant locational factors. This has led to a move from the central city to more convenient and attractive locations that suit the needs of high-tech industry.

CASE STUDY: STOCKLEY PARK, UXBRIDGE, MIDDLESEX

Stockley Park is one of the UK's premier business parks. It is located in London's Metropolitan Green Belt on reclaimed land which was once gravel pits and later a landfill site. As a result, the land had little amenity value and experienced uncontrolled release of methane and leaching, resulting in poor vegetation and pollution which also infiltrated the nearby Grand Union Canal. The value of the site with its close proximity to Heathrow Airport, the M4 and M25, 15 miles west of central London was recognised by the London Borough of Hillingdon in 1980 but it took 4 years before agreement was reached to develop the 400 acres into a business park with 150 acres of buildings and 250 acres of landscaping, including an 18-hole championship golf course and The Arena incorporating shops, restaurants, a health club with 25-metre pool and a central management facility.

This work, commenced in 1986, instigating Europe's largest earthworks programme:

- All 4 million cubic metres of landfill had to be moved north to form the golf course.
- Gravel extracted from the east of the site was used to form the building pads on which the office structures were placed.
- The business park was landscaped and planted to maximise visual and amenity impact.

Research in both Europe and the USA was undertaken to identify the building requirements of potential occupiers. More than 100 companies assisted in this process. It has resulted in the construction of 2 million square feet of office space which is occupied by at least 25 major companies including Marks and Spencer and Cisco Systems, in a high-quality purpose-built environment.

Over 5000 people are employed at Stockley Park. Many job opportunities are made available to local people and Hillingdon benefits from the increased business and domestic rates that are collected from its occupiers. It also benefits from the creation of an environment that has a high amenity value compared to the barren landscape which existed prior to its inception.

Stockley Park also benefits from its close proximity to Brunel University, which is also located in Uxbridge. Brunel specialises in foundation and higher degrees and research into the sciences and computing, both of which are of value to many of the companies based at Stockley Park.

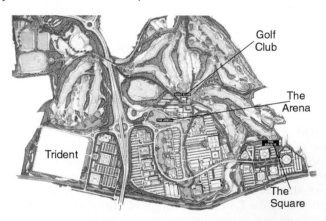

Figure 33 Map of Stockley Park.

Questions

1. Using an area that you have studied, describe the retail hierarchy that exists there.
2. Why is the City of London so attractive to financial and business institutions?
3. **a)** Explain the recent growth in out-of-town shopping centres.
   b) What effect has the growth of out-of-town shopping had on city centres?
4. Using specific examples that you have studied, describe and explain the location of business parks.
5. What are the economic advantages to Liverpool of being granted the title 'European City of Culture, 2008'?

5 Regional Development and the Role of Government

The role of government is to serve the needs of its people and the actions of government with regard to development, may be viewed as a response to the needs of a country as a whole or a region within it. Usually government action can be seen as supportive, but in certain circumstances it may be seen as counter-productive to the well-being of the people.

The examples outlined below aim to illustrate both the positive and negative influences government can have on economic development.

1 UK Government and Regional Development

The dynamic nature of industry means that areas at one time possessing economic advantages for development may later be in decline. This is particularly the case in long-established heavy industries and extractive industries such as coal. The impacts of an industry that was established over a century ago on the development in the mid- to late-twentieth century and beyond were considerable, not only because of a plant closure but also because of the residential fabric which may be left in its wake. Old terraced housing associated with the early phases of the industrial revolution needed replacing and new industry needed to be attracted to replace the old. Such regeneration, with the aid of private finance, was seen as a way of regenerating run-down areas, reintroducing industry with a market for its product and thus reducing the high unemployment inherent in such areas of the city by addressing the skills shortage by also offering retraining.

The British government has been actively intervening in regional development since the time of the 'Great Depression' of 1929 when the first 'assisted areas' were recognised and established to combat the fall in economic activity. Following World War II, 'Comprehensive Development' areas were identified in 1947 that resulted in a slum clearance programme which saw the demolition of 1.5 million houses and their replacement mostly with high-rise apartment blocks. From

1967 to 1977 two new initiatives overseeing housing improvement and social and economic welfare programmes were established. Their work was not concentrated in urban areas as it had been recognised that development in rural areas such as mid-Wales and the Highlands of Scotland was just as important and significant. These policies were carried through irrespective of the political party in power at the time.

The Labour government's White Paper 'Policy for the Inner Cities' in 1977 viewed the inner city as being a problem region requiring specific attention and the Urban Areas Act of 1978 established partnership and programme areas to keep and encourage employment in the inner city. The Conservative's 1979 election victory saw a continuation and expansion of these policies but with a changed emphasis. It saw a public–private partnership with the economic regeneration being the priority over social and environmental issues.

The next 16 years of Conservative control saw the introduction of:

* Urban Development Corporations (UDCs)
* Enterprise Zones (EZs)
* City Action Teams (CATs)
* Inner Cities Initiatives (ICIs).

a) Urban Development Corporations

UDCs, introduced in 1981, were established as partnerships between central government and the private sector of industry. The aim of the partnership was to encourage new industrial and commercial activity in cities with large amounts of derelict and underused buildings. The powers of the UDCs were considerable as they could grant planning permission, acquire, manage and dispose of land and other property, even using the powers of compulsory purchase if considered necessary. They could also give loans and grants, add the infrastructure deemed necessary for property development and advertise.

The UDCs were developed in three phases:

* 1981: London Docklands and Merseyside
* 1987: The Black Country, Cardiff Bay, Teeside, Trafford Park and Tyne & Wear
* 1988: Bristol, Central Manchester, Leeds and Sheffield.

CASE STUDY: LONDON DOCKLANDS

The second UDC in the first phase to be established was in London's Docklands. It was an area of 8.5 square miles of east London in the boroughs of Newham, Southwark and Tower Hamlets which had suffered from economic decline following the closure of the docks as larger container vessels transferred to

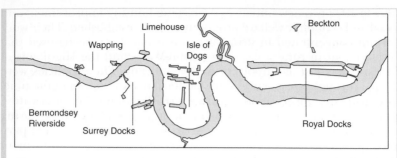

Figure 34 London Docklands UDA.

deepwater docks near the mouth of the River Thames. Between 1978 and 1983 over 12,000 jobs were lost and, as the majority of those made unemployed were blue-collar workers, their skills were not suited to the newer growth industries developing in London.

The Local Government Planning and Land Act 1980 aimed to regenerate the area 'by bringing land and buildings into effective use, encouraging new industry and commerce, creating an attractive environment and ensuring that housing and social facilities are available to encourage people to live and work in the area'.

The London Docklands Development Corporation (LDDC) worked for 17 years and in that time achieved £1.86 billion of public investment and £7.7 billion of private sector investment. As a result:

- 1884 acres of derelict land was reclaimed
- 1066 acres of land was sold for redevelopment
- 144 km of new and roads were added
- The Docklands Light Railway was built giving greater access to the area
- 25 million square feet of commercial and industrial floorspace was constructed
- 2700 businesses were trading in the area
- 85,000 people were working in Docklands
- 24,046 new homes were constructed.

The centrepiece of the work undertaken by the LDDC is Canary Wharf in the West India Docks (see Chapter 4), but other former docks, including Shadwell Basin and Millwall Docks, were also rejuvenated. Once completed, the total public sector cost of the project was £3.9 billion, 48% of which was incurred by the LDDC, 25% by London Transport and 27% by the Isle of Dogs Enterprise Zone. Half of this was spent on new or improved transport infrastructure. By March 1998, £8.7 billion of private sector investment had been invested in Docklands and this investment continues to rise.

In contrast to the LDDC, the Tyne & Wear Development Corporation (TWDC) was responsible for 26 miles of riverside, following the decline in the shipbuilding industry. The corporation worked in a different way to the LDDC by adopting a policy that informed and consulted local people and organisations about their work. It saw its aim to link improvements in the physical environment with wider social and economic regeneration. The results of this enabled the local community to gain from:

- Training and recruitment projects that helped local people find jobs being created in construction and other areas. For example, this enabled 2000 local people to find employment from the Royal Quays Employment Centre.
- A £26 million social housing strategy in partnership with the Housing Corporation that ensured that 25% of the housing in the development area were low-cost homes which were offered for rent or shared ownership.
- Design policies that ensured equal access for people with disabilities.

An evaluation of the work of the TWDC by the Joseph Rowntree Foundation has found that such involvement by the local community has made the corporation's work more effective. This has been backed up by research from the European Institute for Urban Affairs at Liverpool University who also recognise the importance of consultation on such projects producing valuable outcomes for local people.

b) Enterprise Zones

EZs were first introduced by the Conservative government in 1981. The principles of the free-market economy were to operate in the 25 zones created which varied in size from 50 to 440 acres. The aim was to attract growth industries from the commercial and manufacturing sectors of industry to hotspots of high unemployment in urban areas. Sites chosen for enterprise zones included the Isle of Dogs in London, Liverpool–Speke, Scunthorpe and Rotherham, the latter two having experienced large job losses with the closure of steel works in their towns.

The EZs offered incentives for small- and medium-sized industries to locate in them, with:

- exemption from business rates
- exemption from land development tax

- a simplified planning structure
- reduced government bureaucracy
- tax allowances of up to 100% on capital expenditure
- statutory planning controls being administered more quickly.

These concessions lasted for 10 years by which time it was expected that industries could be self-sufficient and the benefit could be passed on to other needy locations. By 1986 it was estimated that 63,000 jobs had been created in EZs.

CASE STUDY: THE ISLE OF DOGS EZ

This was designated in April 1982 on a 195-hectare site including the West India, Millwall and East India Docks, and a small section of the Leamouth area of the London Borough of Newham. Small-scale industrial and commercial developments were built on both sides of Millwall Inner Dock, and the Limehouse Studios were built in a converted warehouse at the eastern end of Canary Wharf. High-density office developments on Heron Quays pre-ceded the 1987 decision for the development of a 29-hectare site at Canary Wharf by Credit Suisse First Boston and later the Canadian developers Olympia & York to match their North American enterprises. Credit Suisse were to use the site as their UK headquarters.

East India dock was also developed with top architects trans-forming the area. The former *Financial Times* and Reuters build-ings together with the relocation from Fleet Street of many newspapers have transformed an area of dereliction into one of high prestige with the Isle of Dogs becoming London's new busi-ness district by 1997.

EZs have, however, been criticised. The estimated cost of £168,000 for each job created and the negative effects on sur-rounding areas caused by companies relocating to sites where savings on indirect taxation could be made led to the designa-tion of no more EZ after 1983.

However, May 2003 has seen the pressure group the Alliance for Regional Aid Campaign (ARAC) propose 39 possible EZs as 'tools for regeneration' as part of a package to stimulate econ-omic activity because they provide a vital impetus to job creation. ARAC are lobbying government to persuade them that a suc-cessor to the EZs, as devised by the Conservative administration, is needed to kick-start development in areas such as Baglan Bay in Port Talbot, South Wales and Deeside in North Wales.

- Baglan Bay is earmarked as a site for the Baglan Energy Park that could attract 6000 jobs, but without additional financial incentives at an early stage full development could take 20 years.
- Deeside is close to the most deprived areas in Flintshire and financial incentives could boost the area's attractiveness for inward investment.

c) City Action Teams, Inner City Initiatives and City Challenge

Nine CATs were established between 1985 and 1987. The aim was to encourage joint development between the public and the private sector. The Inner City Initiative of 1986 created a series of task forces in areas of high unemployment so that development opportunities could be exploited to the benefit of the local community. In 1991 the City Challenge scheme was introduced which gained additional government funding in 1993. It targeted local authorities and other interested parties such as universities and local industry to put forward specific urban projects that would enhance the urban fabric.

City Challenge allocated £37.5 million to each of 31 urban programme authorities over a 5-year period including the Newtown and South Aston district of Birmingham. Birmingham's City Challenge team defined their objectives as being to:

- Stimulate business development, enabling private sector industry to invest for commercial advantage.
- Enhance educational achievement.
- Restore public and business confidence.
- Create a self-sustaining housing market.
- Provide short- to medium-term measures to alleviate poverty.

This development in urban policy was criticised because it indicated a shift in emphasis from local to central control and administration, and an emphasis on private sector investment. It also highlighted the shift in concern from social to purely economic actors. However, the European Institute for Urban Affairs considered it to be the most promising regeneration scheme attempted to that time. It had more influence between the public and private sectors allowing more strategic economic alliances to be formed.

By the time of the fifth year of operation the UK had a Labour administration with their own views on urban policy and their role in regional development.

CASE STUDY: THE LONDON BID FOR THE OLYMPICS, 2012

Government support

In May 2002 the UK government announced that it supported a bid for the Olympic and Paralympic games. The Culture, Media and Sport Secretary, Tessa Jowell, announced that the bid would be 'good for sport, good for London and good for the whole of the United Kingdom'. She said that the cost of staging the games would be £2.4 billion which would be funded from:

- £250 million from the London Development Agency
- £625 million from London council taxes
- £1.5 billion from a new Olympic lottery game and existing lottery funds.

This allows the government to support the bid and avoids additional direct costs being imposed on the tax-paying public.

Benefits of hosting the Olympic Games

The benefits of staging the games at a site in Stratford and the Lower Lee Valley in east London would see a regeneration of an area of low-intensity land use and physical dereliction. East London, or the Thames Gateway, suffers from high unemployment and multiple deprivation, and the siting of Olympic facilities in the area would enhance employment opportunities with 3000 full-time jobs being created and would be a cornerstone of further inward investment.

It is envisaged that a mixed-use commercial and residential development will be developed to house the Olympic Village and that the necessary transport infrastructure will be in place to accommodate:

- 11,000 athletes participating in the 300 events held over a 16-day period
- 5000–6000 coaches and officials
- 4000–5000 members of the 'Olympic family'
- 7000+ sponsors
- 4000 athletes and 2500 officials for the 12-day paralympics.

Accommodation would also be needed in London for:

- 20,000 media representatives
- 63,000 operational personnel
- 500,000 spectators a day travelling around London to view the events.

Up to 100,000 of the 200,000 hotel rooms available in London will be required to accommodate these people and the implementation of the London cross-rail link is considered necessary for the 125,000 spectators who are predicted to cross central London on a daily basis to reach the games. The airport facility, with Heathrow's Terminal 5 complete, and the cross-channel rail link's terminus at Stratford in east London complete are considered to be sufficient to cope with the demand created.

Timetable to decision day

The International Olympic Committee executive board will have decided by June 2004 which of the bids from cities wishing to host the game has been accepted and an evaluation commission survey will announce in May 2005 the shortlist of candidates to go into the final selection. The final decision will be announced in Singapore on 6 July 2005.

Government can be seen to be pro-active in this process and is an example of the private–public partnership common in the UK today.

2 Friedman's Model of Regional Development

In 1964 Friedman created a general theory of urbanisation using the work of Myrdal (Chapter 3) and Hirschman. His premise of regional income differences were the result of differences in the stages of development of different cities.

Stage 1 relates to a pre-industrial society while stage 2 examines the growth of a core settlement to which industry and people are attracted. This later dominates the country as a single metropolitan region. As industry matures in stage 3, peripheral or secondary cores develop to be superseded by a functional interdependent system of cities in stage 4.

In the USA the dominance of the north-east region over all other regions of the country and the ensuing urbanisation associated with industrial development from the nineteenth century onwards led to it becoming the wealthiest area of the country and, according to Friedman, a single national core, a secondary or subcore, has since developed on the west coast in California but the south of the USA remains less urbanised and its people generally in receipt of lower incomes. Friedman considers that the USA is an example of stage 3 of the development model, and that stage 4 can only be reached in countries that are smaller and more compact in size to enable a functionally interdependent system of cities to develop.

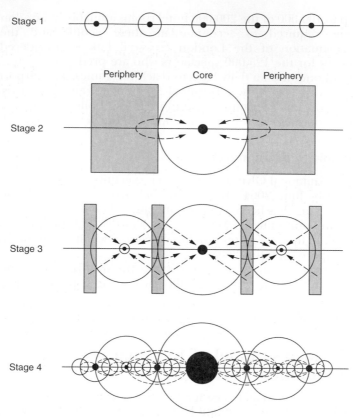

Figure 35 Friedman's development model. *Source*: Nagle (2000).

Friedman believed that government had a fundamental role to play in assisting the growth and development of underperforming regions. He considered it to be a form of cumulative causation as outlined by Myrdal. Poorer areas may gain economically and financially at the expense of already established areas even though the wealthier regions may complain about being drained of their resources to subsidise others.

As Friedman recognised, wealth and affluence are not evenly distributed. Areas may lack the necessary raw materials, workforce or investment to enable wealth to be generated and they may be located away from the core of economic activity. To compensate for the accident of location at the periphery of economic activity, government can initiate policies to stimulate economic activity to benefit the disadvantaged and reduce the financial drain that such areas may have upon the economy of a country as a whole.

This is seen in action in the following case study.

CASE STUDY: THE SCOTTISH HIGHLANDS

The highlands and islands of Scotland form the most northern region of the UK. The rugged terrain and the lack of direct access to many of the northern communities from the south leads to a relative isolation from the dynamism that exists in the central valley of Scotland and other prosperous regions of the UK.

To oversee the needs of the workforce and to ensure that the population exists within the bounds of a viable economy the Highlands and Islands Enterprise initiative (HIE) was formed in 2001 by the Scottish parliament to oversee the economic development and revitalisation of the area.

The HIE is part of a national strategy to deliver to the northern half of Scotland, containing only 8% of its population, economic prosperity. This involves looking at the local communities, their needs and their strengths, and ensuring that in an ever-evolving world a global dimension is in place taking into account the new technologies that are available. A key role for the HIE is to apply for EU funding for the area. To achieve this the agency works in conjunction with the Highlands and Islands Partnership Programme and the Highlands and Islands European Partnership.

The HIE has gone into partnership with Media Lab Europe, the European research partner of the Massachusetts Institute of Technology, to enable businesses and communities to benefit from the new communications and other technologies which are becoming available. This will allow technological innovation to become available to small isolated communities and stimulate economic growth.

This partnership was announced in November 2003 at the same time as research into Scotland's rural economy found that:

- in rural Scotland, the largest occupational group was in skilled trades (16%)
- 14% were self-employed
- 56% of school leavers go on to tertiary education
- unemployment rates are below the national average
- population is expected to decline by 3% by 2016.

By working with Media Lab Europe it is expected that this workforce can benefit from the new technology and build on their skills to assist the prosperity of the region.

One area of potential growth throughout Scotland is tourism and the HIE's local enterprise companies are actively seeking out ways of increasing income from tourism. Orkney Islands Enterprise, for instance, has commissioned a study to assess the potential of boat and sea-shore angling, and concluded that a niche market exists for anglers keen on catching large predatory fish such as conger eels. This would also directly benefit the local hoteliers and restaurants as well as fish suppliers and boat operators.

To improve access to the area, the Highlands Development Agency has provided grants to Eastern Airways to open up new air links to both Manchester and Birmingham, primarily for the business market but also for tourism. This adds to the infrastructural links available to the area and increases access through which further economic benefit can be gained.

It has been the Scottish parliament that has been behind such initiatives and shows how government can directly assist in the industrialisation process in the poorer peripheral regions of a country.

The following case studies highlight how government on a national and local scale can bring about change in both a positive and negative way.

CASE STUDY: ESTONIA – AN EXAMPLE OF POSITIVE GOVERNMENT INTERVENTION

With a land mass of 45,200 square miles, it is the smallest of the three Baltic states of Estonia, Latvia and Lithuania. Since the thirteenth century, Estonia has been ruled by Sweden, Poland and Russia, becoming one of the 15 Soviet Republics making up the USSR from 1944 with only brief periods of independence. Estonia has a population of 1.4 million and was predominantly an agrarian economy before it was incorporated into the USSR after which it became highly urbanised and industrialised.

Post-communism

Reforms in the USSR during the late 1980s brought about a renewed sense of patriotism among the Estonian people and the Estonian Supreme Soviet voted to abolish the Communist Party's dominant role in society in February 1990. A referendum in 1991 resulted in Estonians producing a 77.8% vote in favour of restoring Estonian sovereignty and a failed coup in Moscow by Communist hardliners in August 1991 enabled a full declaration of independence to be made the same month.

The radically different political environment also brought about a radical change in economic policy, from one dominated by central planning and the needs of the Soviet economy to one meeting the needs of the people of the Estonian nation. People even volunteered to pay taxes rather than avoid them as they knew that the proceeds were going towards their own nation rather than the USSR for whom the vast majority of people showed antipathy.

The structure of the economy has changed significantly since 1991. In former Soviet times, oil shale extraction, electricity production, wood processing and the chemical industry were all expanded. Machinery and metal industries were also developed, but not to the same extent as in other republics, and food processing and textiles were also developed. Most industrial production was dependent on the market generated by the USSR itself and its collapse led to a fall of industrial output of 60% between 1992 and 1994.

Economic growth

Growth returned in 1995 as a result of economic reforms allowing privatisation of industry and trade with a broader market. The EU is now a major market for Estonia with Finland, Sweden and Germany accounting for over 52% of the country's merchandise exports in 1999. The remainder of trade is dominated by other EU markets and Latvia and Lithuania, while Russia's market share with Estonia has fallen to only 8%. A similar figure exists for imports from Russia and the re-orientation of trade sees Finland, Sweden and Germany accounting for 57% of imported goods.

Sector	Percentage
Primary industry	7
Manufacturing, including utilities	18
Construction	5
Trade	16
Transport and communication	15
Public services	15
Finance and property	16
Other	8

Figure 36 Estonian gross domestic product (GDP) by activity, 1999.

Estonia's markets are now firmly established in western Europe, and benefits from direct foreign investment from Finland and Sweden in particular whose electrochemical companies are shifting assembly and research facilities to the country. Finnish companies have also taken over many Estonian financial and other institutions, and their common language and cultural ties have been significant in allowing this to happen. This has been of benefit to the Estonian economy. Today, one-third of imports are later re-exported once value has been added to the product. This is also reflected in the forestry sector as instead of unsawn timber being exported, sawmills are now processing the timber into paper, furniture and sawn timber. Sales of these products to the EU have increased by over 30% in 5 years. The importance of the links developed with the EU since gaining independence have been recognised and Estonia joined the EU in May 2004.

The EU

The Estonian economy has changed direction in 10 years, which in economic terms is a very short period of time. New opportunities will be forthcoming with the new markets that will be opened by membership of the 25-nation EU and more diverse economic activity together with a democratically elected government will allow Estonia to evolve and gain in prosperity in its own right rather than as part of a Union dominated by Russia.

CASE STUDY: ZIMBABWE – AN EXAMPLE OF NEGATIVE GOVERNMENT INTERVENTION

Zimbabwe, a landlocked country in southern Africa, covers 390,580 square kilometres of high plateau and high veld. It has a population of 12.5 million [98% African (82% Shona, 14% Ndebele, 2% other), the other 2% being Asian and European], with 40% under 15 years of age and only 3.5% over the age of 65. It is largely an agrarian economy with most of the population living in the rural areas. It also possesses 11 minerals, which are commercially exploited.

Independence and its aftermath

Formerly known as southern Rhodesia, Zimbabwe gained its name on independence from the UK in 1980 following a period of isolation dating back to 1965 as a result of the 'Unilateral Declaration of Independence' (UDI) by the colonial government of Ian Smith. The new government gave the native African people the vote and a stake in their own future that had been denied in the apartheid system previously operated in the country. Robert Mugabe became the nation's first prime minister in 1980 and became president in 1987. He has dominated the political system since independence but he has been accused of rigging the 2002 presidential elections to ensure his re-election. The brutal repression of his opponents, the lack of human rights and the mishandling of the economy have led to widespread criticism from the EU, the USA and the Commonwealth.

Natural wealth

The country possesses a tropical savannah climate with high temperatures throughout the year and seasonal rainfall that can fail. The land varies in fertility, but is capable of supporting the indigenous population and providing the country with surpluses that can be exported. Agriculture was the mainstay of the Zimbabwean economy and in past times it used to supply the United Nations (UN) with food for famine relief operations in other areas of Africa. There is no reason for any Zimbabwean to suffer food shortages or starve, but mid-2003 saw international aid agencies and donors having to feed up to 5 million starving Zimbabweans.

Following independence, the prime farm land in Zimbabwe remained in the hands of the white farmers who formed less than 1% of the population. It was recognised by the government and farmers alike that a land redistribution programme would be needed to redress the balance, but no actions were taken to implement such changes. Most Zimbabwean people remained in their homelands working on subsistence farms with male adults often leaving to work in Harare or Bulawayo where incomes were higher. No effort was made to educate the farmers in how to work their land commercially and the status quo remained.

Land rights and the affect of government policies on the nation

It was only with the rise of a democratic opposition within the country that Mugabe began to address the issue of land rights. His call to provide the army war veterans with land was well received but the means by which he undertook his policy of land redistribution was controversial and led to political acrimony.

The actions of President Mugabe since 1999 have led the country towards financial ruin, food shortages and possible starvation. His policy of 'revolutionary land reform' and the

Spatial Change in the Distribution of Industry on a Global Scale

1 The Cyclical and Dynamic Nature of Industrial Activity

Entrepreneurs, who are the driving force behind any form of industrial activity, are constantly trying to ensure that their industry is at its most efficient and competitive so that shareholders and investors get the best return, or profit, from the enterprise.

To achieve this, investment in current technology and taking advantage of new industrial applications and innovations, as they become available, are important factors which can directly affect not only the location of an enterprise but also its workforce.

a) The Kondratieff cycle

Major technological innovations in the past have led to a shift in the location of industry. This has been well documented with regard to the UK in earlier chapters, but the force of change has been recognised and analysed by a Russian economist Kondratieff who observed the rise and fall of economic output on an approximate 50-year cycle.

His idealised long wave begins with the 'upwave' where prices start to rise slowly and allow economic expansion. At the end of this 25–30-year 'upwave' period, inflation is higher and at its peak brings about the onset of recession that sees a reduction in economic activity and output. The recession is longer and deeper than the expansion that took place in the time of the 'upwave'.

It is only with selective expansion in new industries that new regions or areas of the economy begin to recover. With regard to the UK, four waves can be identified from the beginning of the nineteenth century. Cotton manufacture and the early iron and steel industry signifying the first wave. The second wave of the mid to late nineteenth century saw developments in heavy industry and new technological innovation. The development of the car industry and the chemical and electrical industries in the period before World War II marks the third wave, with the conflict itself accelerating technical innovation and invention. In the post-war period, these technical advances further fuelled the development of the aviation industry and a more sophisticated electronics industry, which were aided by a growth in consumer services. A fifth wave may now be beginning in the UK with the resurgence of car manufacture, electronics, consumer services and the use of information technology.

On a global economic scale, the last cycle or 'upwave' is considered to have begun after World War II, with recession then hitting in the 1980s. Subsequent growth was marred by a further recession in 1990–1991 which meant that the world's economy was on a plateau with only certain areas beginning a new 'upwave'. The UK is on such a wave at the time of writing, while countries such as France, Germany and Japan remain in a state of contraction. The USA, however, recorded its largest growth in over 20 years in December 2003, signalling a return to economic growth. This is based on large tax cuts and the acceptance of a greater budget deficit.

Although Kondratieff's work has received criticism from other economists, it is a worthy study as it does bear resemblance to the growth of the UK and other developed world economies.

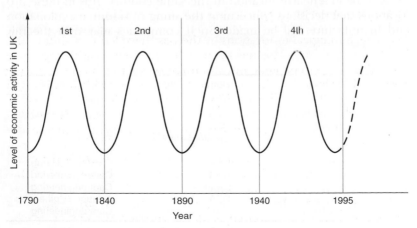

Figure 37 Kondratieff waves.

2 The Role of TNCs and the Spatial Distribution of Industry Worldwide

Economically developed nations, such as the UK, are the driving force behind the world economy and they are also the driving force behind much of the industrial activity that exists in other less developed nations. To fuel their industrial activity, many companies have expanded their operations overseas to take advantage of new markets, cheaper raw materials or have even colonised new territories. This led to the development of transnational or multinational companies, which had precedence in earlier centuries. As industry has progressed, evolved and developed, such companies have needed to invest in and exploit the resources of other nations to remain competitive and profitable.

In the mid-1990s there were at least 37,000 TNCs employing in excess of 73 million people. Of these, 66% were employed in the domestic markets with only 1% of the world's workforce being employed by foreign companies. The top 100 TNCs, however, control 60% of these 37,000 company's assets. They account for in excess of 50% of world trade in manufactured goods and services and 16% of the world's productive assets.

The 10 largest TNCs are shown in Figure 38.

The wealth generated by such companies often exceeds the wealth or GDP of many of the world's smaller economies. Exxon and General Motors, for instance, have sales similar in size to the GDP of Switzerland. This gives TNCs immense economic and political power.

The extent of the operation of a large TNC means that complex planning, organisation and co-ordination are essential for an effective operation. The headquarters are likely to be in a developed world city, usually the one in which the company was first established, and R&D is likely to be in a nearby location in the same country. It is in these two locations that decisions concerning the siting of resource exploitation and manufacture will be made. Such companies may own the raw

Rank	Company	Home country	Industry sector
1	Wal-Mart	USA	Retailing
2	General Motors	USA	Cars/engineering
3	Exxon-Mobil	USA	Oil
4	Royal Dutch/Shell	UK/Netherlands	Oil
5	BP	UK	Oil
6	Ford	USA	Cars/engineering
7	Daimler Chrysler	USA	Cars/engineering
8	Toyota	Japan	Cars/engineering
9	General Electric	USA	Aviation engineering
10	Mitsubishi	Japan	Cars/engineering

Figure 38 The 10 largest corporations, 2003. *Source*: www.fortune.com.

materials, manufacturing and distribution facilities in different countries but decisions about where to manufacture a final product will be made at the company's headquarters in the home country.

Transfers of information and product between plants generates income, taxes and duties for the host nations, and can be seen as effective ways for LEDCs to develop their industrial base. Employment also generates income and taxes for the host, and this has the indirect effect of increasing the political influence of the TNC. However, this will only occur in countries that have resources deemed suitable for exploitation by the transnational. Many LEDC nations will remain underdeveloped unless international trade and aid comes to their assistance.

Multinational companies can, therefore, provide advantages to LEDCs but disadvantages also accrue in the countries in which they operate.

This table shows how multinationals can have considerable influence in the global economy. However, such companies are not immune to the vagaries of the world economy and rationalisation, diversification and re-organisation are commonplace as they adjust to the new innovations often associated with improvements and refinements in information technology.

Advantages	Disadvantages
TNCs introduce new capital, technology, expertise and skills to the country	Capital-intensive investment creates few jobs
The new industries directly benefit the host country as the country may not have had the necessary technology or capital beforehand to develop its energy or mineral resources	Low wages leading to the impression that the workforce is being exploited
Infrastructural improvements, especially transport, are of benefit to local industries and people	Avoidance of local taxes or tax concessions and the export of profit means the country gains little economic benefit
Jobs are created, especially in labour-intensive, light manufacturing	Most products are for export and are liable to changes in world demand and price fluctuation
Exports are increased	Local resources are exploited and production is not based on local needs but their earning potential in a world market
The multiplier effect can stimulate further economic growth	Less stringent health and safety, and pollution controls mean that the workforce and local environment may become adversely affected

Figure 39 The advantages and disadvantages of TNCs.

3 Historical Background to TNCs

The voyages of discovery of the fifteenth and sixteenth centuries and beyond were based largely on the desire to trade and, in particular, to find a sea passage to India and China. At this time, the riches, industrial techniques, and skills of India and China, and Asia in general, were greater and more superior to those of their European counterparts. This was backed by a sophisticated financial infrastructure matching that existing in western Europe at the time.

Companies from Portugal, Spain, The Netherlands and England aimed to develop trade links with the region. In 1599, Queen Elizabeth I granted the English East India Company (EIC) a charter awarding them a monopoly on trade with the east and in the early seventeenth century it attempted to sell British broadcloth to India, the country's most popular export to continental Europe. However, it found no demand for the product but instead found that India possessed many products that could be sold in England.

This led to a huge trade with India and beyond, with commodities such as pepper being imported from Sumatra and Java. Its chief competitor was the Dutch East India Company (VOC), founded in The Netherlands in 1602, which also traded in pepper, spices and other commodities. VOC tried to prevent the EIC from operating in its area of influence – Indonesia. The EIC continued trading with India where it began to acquire land and territories around Bombay, Calcutta and Madras, as well as establishing a military force and base in the region. In 1689 it announced its intention to become a territorial power in India and over the next century established itself as a major power throughout the Indian subcontinent, even repelling French involvement in the region.

Trade expanded into China in the eighteenth century but India remained the dominant area of influence for the EIC. In 1857, the Indian colonies of the EIC became British Colonial India. The wealth generated by involvement in India during this time was used by British banks to invest in industry not only in the UK, but also in Europe and the USA. This funded the growth and spread of the industrial revolution and subsequently the development of the modern multinational company.

Similar policies by VOC secured their wealth and sphere of influence in South-east Asia and it is still felt today. The Netherlands, like the UK, is still home to many of the world's multinational companies.

4 Examples of TNCs: Rio Tinto, Nestlé, Nissan and HSBC

TNCs operate in a wide range of sectors and different ones have been selected as case studies to represent the variety.

a) Rio Tinto (RTZ)

Rio Tinto plc, formerly the Rio Tinto-Zinc Corporation was formed in 1962 by the merger of two British companies: The Rio Tinto Company and The Consolidated Zinc Corporation. Their respective histories go back to 1873 when a company was formed to mine the copper reserves at Rio Tinto in Spain and 1905 when a company was formed to treat zinc-bearing tailings at Broken Hill in New South Wales, Australia.

RTZ has now become a world leader in the locating, extracting and processing of the Earth's mineral resources. Its major mineral exploitations are bauxite (for aluminium), copper, diamonds, gold, iron ore, energy sources such as coal and uranium, and industrial minerals including borax, titanium dioxide, salt, talc and zircon. The company is active worldwide with major concentrations of extraction in Australia and North America, and with interests in South America, Asia, Southern Africa and Europe (see Figure 40).

Chairman Robert Wilson has said that RTZ 'is primarily a mining business [that] only takes the next step into further processing if it

Product	Country	Company
Iron ore	Australia	Hammersley And Robe River Operations
	Canada	The Iron Ore Company of Canada
Coal	Australia	Coal and Allied Industries & Pacific Coal
Coal and copper	USA and Brazil	Kennecott Energy & Minerals
Uranium	Australia	Energy Resources of Australia
	Namibia	Rossing Uranium
Borax	USA	Borax & Luzenac Operations
	Europe	Borax & Luzenac Operations
Titanium	Canada	Rio Tinto Iron & Titanium
	South Africa	Rio Tinto Iron & Titanium
Salt	Australia	Dampier Salt
Aluminium	Australia and New Zealand	Comalco
	Europe	Comalco
	Guinea	Comalco
Copper	Portugal	Neves Corvo
	Australia	Northparkes
	South Africa	Palabora
	Sweden	Zinkgruvan
	Chile	Escondida
	Indonesia	Freeport/Grasberg
Diamonds	Australia	Argyle Diamonds (Australia)
	Canada	Diavik Diamonds
	Zimbabwe	Murowa Project

Figure 40 The locations of selected RTZ global operations.

clears our financial hurdles, enhances the value of the relevant mining operations and enables us to establish a long term competitive advantage'. Investments are selective and are dependent on the opportunity a potential investment may have to enhance shareholder value. In such commodities as minerals, companies have little control over price so RTZ focus their investment on efficiency and costs of production as the foundation for their profit base.

RTZ announced in December 2003 that its Hamersley subsidiary would increase production of iron ore at its Yandicoogina mine in Western Australia from 20 to 36 million tonnes at a cost of $920 million (£530 million) to meet demand from China. Hammersley, which ranks second in the world in terms of iron ore production, is benefiting from the growth in demand for iron ore from China who is likely to become the world's largest importer, overtaking Japan in this market in 2004.

Other operations are undertaken by wholly owned subsidiary companies such as Borax and Pacific Coal, partly owned subsidiaries such as Palabora, and non-managed associate companies such as Neves Corvo in which governments, public shareholders and other companies have a financial interest.

A criticism often levelled at TNCs is that they ignore environmental issues and take advantage of the cheaper workforce found in many LEDCs.

Figure 41 RTZ mining for gold in Papua New Guinea.

RTZ aims to minimise the adverse effects that mining operations can have on the environment. At the Kelian Equatorial Mining operation in Indonesia, RTZ's largest primary gold mine was opened in 1992, and 500,000 ounces of gold are produced annually. Rigorous controls of all by-products of the mining operation were introduced. Tailings, the remains of the processed ore once the gold has been extracted, are disposed of in the opencast pit once it has become exhausted. This is to contain the acid rock drainage and prevent pollution. The stockpiles of gold-bearing ore are covered prior to processing and dams are constructed to store the acidified water. This can later by filtered before being released into the natural environment.

RTZ aims to work within strict guidelines by contributing to sustainable development and by ensuring that health and safety issues are prioritised. The Lelian operation was awarded a four-star safety award in the year 2000 from the National Occupational Safety Association of Australia. Local people are employed at all levels, with only 45 of the 800 employed being expatriates. They ensure that the benefits of their operations are felt by the local people by providing infrastructure, schooling (including scholarships), health care and training in agricultural activities.

The Lelian operation is due to close in 2004 as the ore reserves will be exhausted, but RTZ has assisted in a development plan to ensure the area's prosperity once the mine has closed.

Responsibility to the workforce

RTZ have also developed and implemented a global HIV/AIDS strategy as a response to the social and economic effects of the epidemic. This follows the guidelines laid down by the International Labour Organisation (ILO) Code of Good Practice on HIV/AIDS and has seen the company initiate a fourfold programme which is based on:

- prevention, awareness and education
- voluntary counselling and testing
- wellness and treatment
- monitoring and evaluation of the company's service provision.

RTZ's mission statement and view of their operation is based 'on our shared commitment to be a dependable global partner and good local neighbour' (www.riotinto.co.uk). Their operation is somewhat different to other TNCs in that it is not reliant on the cheapest labour force to ensure the success of its operation.

b) Nestlé

Nestlé was founded in 1866 and has since become the largest food and beverage company in the world. In 2001, sales amounted to 84.7 billion Swiss francs (SF), generating a profit of SF6.7 billion for that year. Nestlé employs 254,099 and they operate 520 factories in 82 countries. Their headquarters are at Vevey in Switzerland.

Origins of the company

Henri Nestlé, a trained pharmacist, produced a source of infant nutrition that was later marketed throughout Europe under the Nestlé brand, along with a condensed milk product that was launched in 1875. In the same year, Daniel Peter devised a way of combining milk and cocoa powder to create milk chocolate and Peter's company merged with Nestlé in 1904 making it a world leader in such products. With the addition of Maggi & Company, producing soups, sauces and flavourings, it broadened the range of products made by Nestlé and a further merger with the Anglo-Swiss Milk Company in 1905, established the company as a major force with factories in Europe and North America.

Manufacturing began in Australia in 1907, and warehouses were built in Singapore, Hong Kong and Bombay, India to meet the Asian demand for the company's products. World War I caused shortage problems but the demand for dairy produce meant that the company doubled its production during the war years and had 40 operative factories.

Post-World War II

The post-war years saw R&D leading to the production of the world's first soluble coffee powder, Nescafé. The onset of World War II saw Nestlé's sale increase by 125%, and the post-war period saw Nestlé consolidate these operations and take on new acquisitions in frozen food and tinned produce. This was later followed by diversification into cosmetics, becoming a major shareholder in L'Oreal in 1974.

The oil crisis of the mid-1970s saw the quadrupling of the price of coffee and deteriorating exchange rates against the Swiss franc led to further diversification with the acquisition of Alcan Laboratories Inc., a US manufacturer of pharmaceutical and ophthalmic products. The US food company Carnation was purchased in 1984 and further diversification has been seen with the purchase of Spillers PetFoods in the UK in 1998, Dreyer's Ice Cream, Chef America and the Ralston Purina PetCare businesses in 2002.

By 2003 Nestlé was involved in the manufacture, processing and retailing of the following products:

- coffee
- water
- other beverages
- dairy products
- breakfast cereals
- infant foods
- performance nutrition products
- health care nutrition products

- culinary products
- frozen foods
- ice cream
- refrigerated products
- chocolate and confectionary produce
- food services and professional products
- petcare produce
- pharmaceutical products
- cosmetics.

Worldwide sales of the products of the Nestlé brand in the period 2000–2002 are shown in Figure 42.

Nestlé's products are primarily agricultural in origin. The company has developed a rigorous environmental policy to ensure that their operations are sustainable and safe. The manufacturing processes operated in the company's factories are designed to be eco-efficient by minimising resource consumption, and reducing waste and emissions while maximising production. Environmental performance indicators, applied to the manufacturing operations outlined in 1997, have shown a fall in overall water consumption in the company's factories by up to 25%, carbon dioxide emissions have fallen by 18% and energy efficiency has increased by 20%.

Such actions enable the company to be seen as a responsible multinational company and Nestlé also sees that its workers are not exploited.

Region	2000	2001	2002
Europe	26,285	26,742	28,678
Western Europe	24,546	24,655	26,424
Eastern and Central	1,739	2,087	2,254
Americas	25,524	26,598	29,293
USA and Canada	15,035	15,548	19,425
Latin America and the Caribbean	10,489	11,050	9,868
Rest of the world	15,710	15,458	14,880
Oceania and Japan	5,606	4,998	4,757
Other Asian markets	5,955	6,380	6,234
Africa and the Middle East	4,149	4,080	3,889
Nestlé waters	5,947	7,418	7,720
Europe	–	–	3,504
USA and Canada	–	–	3,739
Others	–	–	477
Other activities (mainly pharmaceutical)	7,956	8,482	8,589
Total group	81,422	84,698	89,160

Figure 42 Sales of Nestlé products, 2000–2002 in SF millions.

Responsibility to the workforce

Policies in place for their workforces around the world include:

- The Philippines – the starting hourly wage is 2.5 times higher than the legal minimum wage.
- Mexico – support programmes to ensure workers complete their basic and secondary studies.
- Chile – financial assistance to help workers acquire their own housing.
- Japan – working mother programmes including shorter working hours, assistance with child care and a system for future re-employment.
- Brazil – co-operation with AACD, the Brazilian institution for the disabled, ensuring that people with disabilities can be successfully employed.
- USA – ethnic diversity initiatives ensuring that the Afro-Caribbean and Hispanic communities are not discriminated against in employment.
- South Africa – support of EcoLink, a non-government organisation, enabling 150,000 people in local communities to gain access to sources of clean drinking water and plots of land to cultivate vegetables to improve diet.

Sustainability and the development of new products and markets are at the heart of Nestlé's future development. As the world moves out of the global recession, Nestlé, like other transnational companies, will aim to build on its experience and success in the market place.

c) Nissan

The company

The Jidosha-Seizo Kabushiki-Kaisha or Automobile Manufacturing Company Limited of Japan was first established in December 1933. It took over the responsibility for the manufacture of Datsun Cars from the Tobata Casting Company Limited and it became Nissan Cars on 1 June 1934.

Nissan was typical of many Japanese companies that saw rapid modernisation of its industry at that time. The company began exporting cars to Australia in 1935 and in 1936 purchased, from Graham-Paige Motors Corporation in the USA, design plans and plant facilities to further its production. However, the company changed its production from cars to military trucks as World War II approached.

Post-World War II

It was during the post-war era that Nissan once again returned to car production at its Yokohama plant, producing Datsun trucks from 1945 and cars from 1947. In 1952, it began producing Austin cars following a technical agreement with the UK's Austin Motor Company

Limited. Later using its own designs, Datsun became a major car producer in its own right. In 1959, Datsun opened its first overseas plant in Taiwan. The Yulon Motor Company Limited assembled kits imported from Japan. This was followed by a similar investment in Mexico in 1961 with the creation of Nissan Mexicana. Datsun began exporting its own product on a large scale in the early 1970s with its Sunny model, which was extremely fuel efficient, leading the export drive to both the USA and Europe.

At this time US cars, in particular, lacked fuel efficiency and the success of the Sunny led to calls for import quotas on this and similar models from other Japanese manufacturers and the opening of local production plants in the USA. This led to the Nissan Motor Manufacturing Corporation, USA being established in 1980 with the first cars being produced in June 1983. This was followed by Nissan Manufacturing (UK) Limited in 1984.

Nissan has invested greatly in car manufacture in the UK in the past 20 years as a means of ensuring access to the EU market. A state-of-the-art manufacturing plant was opened in 1984 at Washington, Tyne & Wear in north-east England on the site of the former Sunderland Aerodrome with £100 million of assistance from the government. By 2001 the plant was producing 350,000 cars a year that was to increase to 500,000 after a further £235 million investment by the company. At present 5000 employees produce three models of car that are built to cater for the whole of the European market.

Initially, many of the components for the cars were imported from Japan and the Far East, but Nissan has since encouraged local component companies to bid for and provide the components needed for these vehicles, thus making the cars more European in their origin. Nissan even exports some of the cars back to Japan in a complete reversal of earlier operations.

Concern has been expressed about the UK's position outside of the single European currency (euro) and what affect that this could have on investment by the company. The worry over the UK being outside of the euro has even extended into the supply of component parts, up to 65% of which are being billed in euros rather than pounds. The present strength of the pound against the euro may mitigate against the production of component parts in the UK, especially as the Sunderland plant had agreed to reduce costs of production by 30% to secure the new investment.

To ensure access to the most important markets in the world, Nissan has had to spread production across the globe rather than concentrate production at one plant in Japan. Today, Nissan has manufacturing and assembly plants in 17 countries. This globalisation of the company has led to the establishment of regional headquarters in the USA and Europe to oversee the company's operations in these regions. In Japan, Nissan now has plants at Yokohama, Kyushu and Iwaki producing cars using advanced technology.

	November 2003 (units)	Year-on-year change (%)	April–November 2003 (units)	Year-on-year change (%)
Passenger cars, Japan	98,475	−10.6	810,900	+1.6
Commercial vehicles, Japan	18,923	+20.8	159,409	+17.1
Total production in Japan	117,398	−6.7	970,309	+3.8
USA	49,981	+52.2	381,595	+38.7
Mexico	25,175	−9.1	208,647	−14.2
UK	27,943	+39.7	226,534	+10.5
Spain	10,319	+5.1	78,754	+25.6
Others	19,791	+11.4	137,295	+14.4
Overseas production	133,209	+23.2	1,032,825	+14.0
Global production	250,607	+6.5	2,003,134	+8.8

Figure 43 Global production of Nissan cars and commercial vehicles, April to November, 2003. (The 'others' category includes Taiwan, Thailand, The Philippines, South Africa, Indonesia and China.) *Source*: www.nissan-global.com.

Nissan has also gone into partnership with other car manufacturers. In March 1999, Nissan and Renault of France signed an agreement that initiated a global alliance aimed at achieving profitable growth for both companies. This partnership even enabled Renault to use Nissan's showrooms in Australia to gain access to the Australian market.

Nissan's 1999 Revival Plan aimed to ensure the company's long-term global profit and this was achieved in 2001, a year ahead of schedule.

Production in the 6-month period from April to November 2003 is shown in Figure 43.

Nissan aims to increase global sales by a million units by 2006 and to do this it will need to respond to the new emerging markets, especially as the home market in Japan remains in a state of recession.

The growth of car ownership in India and China is likely to see Nissan further exploiting these markets by establishing production facilities as a way of overcoming import tariffs and quotas as it did in both the USA and Europe.

d) HSBC

Since incorporating the former Midland Bank, the Hong Kong and Shanghai Banking Corporation (HSBC) has become the biggest of the UK's major high street banks. Initially established in 1865 to finance the growing trade between China and Europe, it established a network of agencies in China and South-east Asia with further representation on the Indian subcontinent, Japan, Europe and North America during the remainder of the century and into the twentieth century. Trade in bullion, exchange and merchant banking were significant areas of activity at this time and after World War II it played a key role in the reconstruction of Hong Kong.

Its two headquarters are in Hong Kong and London, and its network consists of over 9500 offices in 79 countries throughout the world. It now exploits high-tech innovations to ensure that it operates at maximum efficiency and produces the best dividends for its 200,000 shareholders in over 100 countries. To achieve this, it constantly monitors and alters its business operation to maximise its assets.

This has led to a shift in location of aspects of its operation. October 2003 saw the announcement of the closure of service centres in Sheffield, Birmingham, Brentwood and Swansea in the UK and their transfer to Hyderabad in India. This outsourcing of employment will see 4000 jobs being lost in the UK where costs are 40% higher due to higher wages and increases in national insurance and pension provision in 2003.

Bill Dalton, chief executive of HSBC's UK operation, said that such outsourcing was the only way to guarantee job security for bank personnel around the world and that the bank, 'as one of the world's largest financial companies, has a responsibility to all of its stakeholders to remain efficient and competitive'.

HSBC is not alone in transferring call-centre work to India. British Telecom, Lloyds TSB, Abbey, British Airways and Barclays are some of the blue-chip companies all taking advantage of the lower cost base and educated workforce available in India.

Senior industry figures in India expect technology-enabled services to grow by 65% a year in future, fuelled by the 1.5 million graduates who qualify per year and are also fluent in English. Bangalore has become the centre of the information technology development in India.

TNCs will constantly strive to maintain profitability and this factor alone will ensure that the dynamic nature of industrial activity will continue with aspects of their businesses being moved to the most advantageous locations. Such mobility will continue to be challenged

Company	Number of jobs lost
Lloyds TSB	1500
AXA	800
Aviva	1000
British Telecom	2200
Ebookers	600
Powergen	300
British Airways	1000
Tesco	350

Figure 44 Announced job losses.

by workers' organisations but the power of such companies is such that the most cost-effective labour force will be used in areas where high-tech information technology applications in particular can be utilised.

TNCs are major forces in the world economy. They operate in all areas of economic activity from primary, to secondary and tertiary industry. Their integrated approach means that often such companies are the extractor, processor and seller of the product. Such complex business activity requires vast quantities of capital to manage their worldwide assets. TNCs are often accused of assisting the globalisation of world trade and their power is often criticised. The location of their headquarters in MEDC cities, while their sources of wealth lie in the exploitation of the resources in the LEDC nations is the root cause of this criticism. Companies such as RTZ and Nestlé are aware of such criticisms and respond by being seen as responsible and sympathetic to the needs of their employees and the environments in which they operate. Other TNCs are endeavouring to match this approach as the power and influence of such companies grows.

Questions

1. Explain the term 'Kondratieff waves' and identify the four economic waves of the UK economy.
2. What were the early origins of the current TNCs?
3. With reference to one example you have studied, describe the structure and geographical distribution of a TNC.
4. Why do some large companies decide to become transnational?
5. Describe the impact that TNCs can have when they establish operations within LEDCs.

7 Where Are We Going?

The dynamics of the world economy have altered due to changes in industrial dominance within the twentieth century. The initial advantages won by the UK as instigator of the 'industrial revolution' in the nineteenth century were later lost to other European and US competitors. The industrial advantage moved to the NICs of the Far East. The pendulum has partly swung back in favour of the UK but new industrial powers are rising with the economies of China and India now gaining momentum.

1 The Level of Economic Development in 2003–2004

In 2003 Japan was in recession. The first signs of recovery in the economy were led by rises in exports at the beginning of 2004. Meanwhile, the UK and the USA were both seen to be gaining renewed strengths in their economies. The UK had seen 12 years of growth, while the US economy is being assisted by tax cuts and renewed industrial activity as a result of the Iraq conflict. Germany remains in economic recession and France experiences similar economic problems with falling output and high unemployment.

Even in recession these countries still dominate the world economy and account for a disproportionate amount of the world's GDP and the GNI (gross national income), as the following 2001 figures show:

- world average GNI is $7160
- MEDC average GNI is $22,030
- LEDCs average GNI is $3360
- excluding China, the GNI for LEDCs is $3570.

The fastest growing and emerging economies are Brazil, Russia, India and China (BRIC), according to a 50-year projection

Continent	GNI per capita (PPP*)
Africa	2,120
Asia	4,290
Europe	16,270
North America	30,405
South America	6,820
Oceania	18,400

Figure 45 GNI PPP per capita by continent for 2001. *US dollars at purchasing power parity. *Source*: Population Reference Bureau.

commissioned by the investment bank Goldman Sachs. In a paper by Dominic Wilson and Roopa Purushothaman entitled *Dreaming with BRICs: The Path to 2050*, projections are made using a growth model incorporating factors such as technical progress and population change.

By continent, the GNI indicates even greater extremes – see Figure 45.

The values in Figure 45 are averages for groups of countries and mask extremes within them. It is only in North America, consisting of Canada ($26,530 GNI) and the USA ($34,280 GNI), where there are two nations to consider, both of which belong to the group of MEDCs dominating the world economy.

2 Comparing Continents

a) Africa

An overview

Africa is the poorest of the continents, but there are great variations within it. North Africa's average figure is $3600, but it is sub-Saharan Africa, whose average GNI figure is only $1710, that suffers the greatest deprivation. South Africa's GNI of over $10,000 reflects its more advanced economy. Neighbouring countries like Botswana and Namibia also benefit from having high GNI due to South Africa's regional economic power.

The north African nations of Egypt, Morocco and Tunisia benefit from tourism, while Algeria benefits from oil and this boosts the average GNIs of these nations. The return of Libya to the international fold, with its rejection of weapons of mass destruction and the settlement of claims against its apparent attacks on US and French civilian airliners, is likely to see the nation developing its oil reserves and giving its people increased economic well-being. Other North African nations experience similar problems to those of its sub-Saharan neighbours resulting from a lack of marketable raw materials.

Country	GNI per capita (PPP*)	Adult literacy rate
Botswana	7,410	79.8%
Democratic Republic of Congo	630	65.5%
Egypt	3,560	57.7%
Malawi	560	62.7%
Mali	770	46.4%
Sierra Leone	460	31.4%
South Africa	10,910	86.4%
Zambia	750	80.6%

Figure 46 GNI (PPP) in selected African nations for 2001.
*US dollars at purchasing power parity. *Source*: Population Reference
Bureau and *CIA Word Factbook*.

Botswana

Botswana has one of the higher GNIs in Africa and has seen considerable growth in its economy since independence in 1966. This has been the result of fiscal discipline and management of the economy allowing the country to transform itself from one of the continent's poorest countries. Much of the growth can be attributed to the expansion of diamond mining within Botswana that accounts for one-third of the country's wealth and 90% of the country's exports. Subsistence farming, cattle ranching and tourism are also key sectors of the economy. However, the country still has to deal with an official unemployment rate of 21%. Its effect has been to place a large percentage of Botswana's population in poverty and restrict these potential wage earners access to the wealth of the nation.

Namibia and Malawi

Namibia is similar to Botswana in that it is a prime exporter of extracted and processed minerals including diamonds, uranium and zinc. It too has considerable inequality in the distribution of income in the country, with 33% of its people having an annual GNI of under $1400. This figure, however, is still higher than many other sub-Saharan countries. Malawi's predominantly agricultural economy, with approximately 90% of the population living in rural areas and agricultural activity accounting for 40% of the country's wealth, has hindered economic development. A lack of minerals or raw materials required by the MEDC's has led to a dependence on aid from the International Monetary Fund (IMF), the World Bank and donor nations.

South Africa

South Africa, with its abundant supply of minerals and natural resources, also possesses an advanced and modern infrastructure, enabling the distribution of goods throughout the country. It has a well-developed financial sector with one of the 10 largest stock exchanges in the world and an efficient energy sector. Its workforce is largely employed in secondary manufacture (29%) and tertiary activity (66%), and is different to most of Africa where the primary sector dominates economic activity. However, South Africa does still suffer from high unemployment and the needs of the disadvantaged people living in many of the townships still have to be addressed.

b) Europe

An overview
In contrast to Africa, Europe is the second wealthiest area of the world after North America. Northern and western Europe are the most affluent with the southern and eastern European nations being the poorest. The post-communist nations of eastern Europe have lower GNIs, but the addition of countries including Poland, Estonia and Latvia to the EU in May 2004 will assist in their economic development as new markets are opened up for their industries and inward investment is likely to be increased. The economies of southern Europe were mostly agriculturally dominated, but now possess strong secondary and tertiary sectors. Northern and western Europe, consisting of countries such as Austria, Denmark, France, Germany, The Netherlands, Norway, Sweden, Switzerland and the UK are all highly industrialised and developed, and this is reflected in higher GNIs.

Austria
Austria has a market economy and is a member of the EU. Economic growth between 2000 and 2003 averaged 0.9% due to its links with Germany – 31.5% of Austrian exports were to Germany and 42.6% of imports in 2002 were from Germany.

Germany
Germany became a unified nation in 1990 and the former West Germany had to invest considerable funds to develop East Germany's productivity and wage levels up to Western standards. This cost

Country	GNI per capita (PPP*)	Adult literacy rate
Austria	26,380	98%
Estonia	9,650	99.8%
France	24,080	99%
Germany	25,240	99%
Greece	17,520	98.6%
Italy	24,530	99%
Poland	9,370	99.8%
UK	24,340	99%

Figure 47 GNI (PPP) in selected European nations for 2001.
*US dollars at purchasing power parity. *Source*: Population Reference Bureau and *CIA World Factbook*.

approximately $70 billion. This strained Germany's highly developed economy even though it still remains the third wealthiest nation in the world. Growth in Germany in 2002 and 2003 fell by 1%, and Germany broke the EUs budget rules by going beyond the 3% deficit rule, whereby a country's debt can only amount to 3% of a country's GDP, for which it is to be fined by the European Commissioners.

Poland
Poland joined the EU on 1 May 2004. This former communist nation followed a policy of liberalising its economy throughout the 1990s but the economy is still dominated by state employment and ownership of many of its key sectors. Membership of the EU has influenced recent activity within the country, in particular the importance of trade with Germany, which accounts for over 30% of Poland's trade.

UK
The UK is one of the four trillion-dollar economies of Europe. It has an efficient agricultural system producing 60% of the country's food needs from 1% of its workforce; primary mineral production industry that accounts for 10% of GDP; an efficient manufacturing industry; and an expanding tertiary sector employing 73.3% of its workforce and making it the fourth wealthiest nation in the world.

Between these two extremes in economic activity and development are found the remaining continents whose fortunes lie between the two extremes of Africa and Europe.

c) Asia

An overview
The 51 nations of Asia include oil-rich nations such as Saudi Arabia and Kuwait; former Soviet republics such as Turkmenistan and Uzbekistan; emerging economies such as China, South Korea and Taiwan; and countries as diverse as India and Pakistan, and Japan, the only MEDC nation. Varying GNIs and levels of economic development exist within these countries.

China
China is presently experiencing an annual growth rate of 8% and, according to a report by an investment bank, Goldman Sachs, the economy is expected to outstrip the French economy in 2004, the UK economy by 2005 and that of the USA by 2041, as it increases its manufacturing industries. In 2002, China replaced the UK as the world's fifth largest exporter. It also attracts increasing amounts of inward investment. This currently totals $53 billion compared to the UK's $26 billion in the same period.

The long-term affect of this will not see China's GNI or GDP outperforming that of the UK or USA as their lower cost base will keep these figures at least 50% below the UK's and 40% below the comparative figures for the USA.

India

India has the second largest population of any country in the world. It possesses a diverse economy with traditional subsistence farming, commercial agriculture, modern industries and a growing series of support services. The growth of the car and motorcycle manufacture in India benefits from its lower cost base with wages accounting for as little as 3% of production costs compared to 10% in Europe, for example. This led to MG Rover choosing to import, rather than develop, a new small car from Tata where production-line workers earn 12,000 rupees or £155 a month. Over 16,000 City Rovers, a development of Tata's Indica car, will be imported annually.

Similarly, the move of many UK call centres and associated service industries to India is largely due to the educated workforce fluent in English. Software services form much of the tertiary sector in India and at current estimates India's economy will outstrip the economy of the UK by 2050. This will not mean that the GNI will be similar per capita as the size of the population and the lower wage base will not enable this to occur, but its effect will be to increase the wealth and standard of living for many of the people. Much of this development is due to increased foreign investment following on from a relaxation by the Indian government on controls on inward investment into the country. Great progress has been made in reducing poverty in the country but 440 million people are still surviving on an income of under $1 a day.

In contrast to China and India, the economies of Turkmenistan and Uzbekistan are experiencing the difficulties of adapting to the break up of the Soviet Union and the command economy.

Country	GNI per capita (PPP*)	Adult literacy rate
China	3,950	86%
India	2,820	59.5%
Japan	25,550	99%
Kuwait	21,530	83.5%
Pakistan	1,860	45.7%
Saudi Arabia	13,290	78.8%
South Korea	15,060	98.1%
Thailand	6,230	96%
Turkmenistan	4,240	98%
Uzbekistan	2,410	99.3%

Figure 48 GNI (PPP) in selected Asian nations for 2001. *US dollars at purchasing power parity. *Source*: Population Reference Bureau and *CIA World Factbook*.

Turkmenistan

Turkmenistan became an independent republic in 1991 following the break up of the Soviet Union. Its government is authoritarian and the country has a tribally based social structure. The former Soviet economy is inefficient and the country aims to use its reserves of natural gas and ability to produce cotton to assist its economic development and reduce external debt. There was a 38% rise in exports in 2003, largely due to the international rise in the value of oil and gas prices, but Turkmenistan still experiences widespread poverty, high foreign debt and a lack of market reforms that would induce foreign investment.

Uzbekistan

Uzbekistan, which also gained independence from the Soviet Union in December 1991, initially supported the economy with financial subsidies and controls on production and prices. It has an authoritarian government with a president and executive branch that rules the country. Problems still exist with this rigid control by central government but the country, now the second largest producer of cotton in the world and with significant gold, natural gas and oil reserves, has become a net exporter of commodities. Since independence, there has been a major increase in the disparity between the well paid and the poorer elements of society, with the latter gaining little from the country's independence.

d) Latin America and the Caribbean

An overview

Contrasts in economic development within Latin America and the Caribbean are considerable. Many countries possess strong economies while others are weak.

Argentina

Argentina has a broad primary and secondary industrial base and a literate population, but in recent years it has experienced high inflation, external debt, budget deficits and a loss of confidence in its economy, which has led to a withdrawal of capital investment from the country. The economy began to recover in 2003 and unemployment has begun to fall, output increased by 5.5% and inflation has been reduced by 4.2%.

Brazil

Brazil has the largest of all Latin American economies and a strong primary, secondary and tertiary sector of industry. Assistance from the IMF in the form of a $41.5 billion support programme in November 1998 allowed Brazil to adjust its fiscal or monetary programme and enable structural reform of its economy to take place. This has

Country	GNI per capita (PPP*)	Adult literacy rate
Argentina	10,980	97.1%
Brazil	7,070	86.4%
Ecuador	2,960	92.5%
Haiti	1,870	52.9%
Mexico	8,240	92.2%
Peru	4,470	90.9%
Trinidad and Tobago	8,620	98.6%

Figure 49 GNI (PPP) in selected Latin American and Caribbean nations for 2001. *US dollars at purchasing power parity. *Source*: Population Reference Bureau and *CIA World Factbook*.

allowed GNI to increase at a rate of just under 2%, but the long-term outlook for Brazil is that it will be one of the four strongest emerging economies of the next 50 years.

Haiti
Haiti is dominated by subsistence agriculture and 80% of its population live in abject poverty. Its economy shrank by 0.9% in 2002. Democracy is weak and elections in May 2000, which saw President Aristide re-elected, were considered by the international monitors to be fraught with irregularities. This led to the withdrawal of most international aid from the USA and the EU totalling $500 million by the beginning of 2003.

Mexico
Mexico has a mixed economy with both modern and outdated industry and agriculture. Its proximity to the USA and its trade with that country has had a negative effect on its economy with its slow-down reflecting that of the USA. Industry is dominated by the private sector and, although a free-market economy, income distribution is unequal with extremes of wealth and poverty being seen within the country.

e) North America

An overview
North America is the wealthiest continent in the world and is dominated by the USA.

Canada
Canada has an economy resembling that of its southern neighbour, the USA. Canada is rich in natural resources such as diamonds, gold and iron ore, and has developed a sophisticated mining, manufacturing and service sector to compliment its agricultural economy, which

Country	GNI per capita (PPP*)	Adult literacy rate
Canada	26,530	97%
USA	34,280	97%

Figure 50 GNI (PPP) in North America for 2001. *US dollars at purchasing power parity. *Source*: Population Reference Bureau and *CIA World Factbook*.

was its initial source of prosperity. It is affluent, highly industrialised and market oriented. Like all MEDCs, Canada's population is largely urban and employed in value-adding industries. Proximity to the USA has recently had a negative effect on its economy in a similar way to Mexico, with real growth falling in 2001 but a moderate recovery was seen in 2002. Its abundance of natural resources, a skilled workforce and investment in infrastructure has put Canada in a strong position for the future and a return to growth.

USA

The USA possesses the largest and most technologically powerful economy in the world and is based on a free market economy which provides the population with the highest average GNI in the world. The USA is a world leader in medical, aviation and military equipment, and also dominates in the computing industry. There does exist, however, a two-tier labour market with those lacking educational and professional or technological skills failing to get the same pay awards, health insurance coverage and benefits as the more advantaged. Since 1975 most gains in household income have gone to the top 20% of households, exaggerating the differences between rich and poor. The economy of the USA is resilient and is currently coming out of a period of stagnation and decline. Its low level of government intervention compared to other MEDCs has enabled the country to prosper in the past. However, there is a need for further government investment in the country's economic infrastructure. Similarly, the effects of an ageing population on medical and pension costs, together with low pay among the poorest groups of its population and a massive trade deficit, all need to be managed by the Federal government in both the short and long term.

f) Oceania

An overview
Oceania encompasses the island states of the Pacific Ocean, the continent of Australia and the islands of New Zealand. Many of the island chains are isolated and their economies are small, relying on farming and fishing. In contrast, Australia and New Zealand are classed as

first-world nations on a par with Europe and North America. This dates to the British colonisation of these lands and the exploitation of the resources and the later development of British-style constitutions, economies and levels of development.

Australia

Australia's economy is similar to the top four European economies based on GNI. The wealth is due to the exploitation of the vast mineral reserves possessed by the country, and the utilisation of these to develop a manufacturing base that serves its home and export market. The recent global slump in economic activity has been offset by increases in demand in the domestic economy. Australia has the strongest economy in Oceania.

Fiji

Fiji is one of the most developed Pacific economies based on the exploitation of forests, fisheries and minerals. The processing of sugar cane accounts for 33% of Fiji's industrial activity. The growing tourist trade also give the country important foreign exchange reserves, but the country needs greater inward investment, a managed government budget and to settle land ownership rights between native Fijians and the Indian community which first settled in Fiji during nineteenth century British colonial rule.

New Zealand

New Zealand originally depended on agricultural trade with the UK where it gained concessionary access to the market. Following the UK's joining of the EU and the removal of this advantage, New Zealand has had to develop a more industrialised and free-market economy to compete in the global market. The industrial sector of the economy has developed and a broader agricultural market is now served, but New Zealand still lags behind the GNIs of the major European nations and growth in 2003 was 2.5%.

Country	GNI per capita (PPP*)	Adult literacy rate
Australia	24,630	100%
Fiji	4,920	93.7%
French Polynesia	5,000	98%
New Caledonia	14,000	91%
New Zealand	18,250	99%
Samoa	6,130	99.7%
Vanutu	3,110	53%

Figure 51 GNI (PPP) in selected Oceanic nations for 2001.
*US dollars at purchasing power parity. *Source*: Population
Reference Bureau and *CIA World Factbook*.

Vanuatu

Vanuatu has an economy based on subsistence agriculture that employs 65% of the population. Fishing, financial services and tourism are the other areas of significance in Vanuatu's economy. It has a lack of mineral resources that, together with Vanuatu's isolation and susceptibility to natural hazards, have left the country under-developed, and with a fall in national income of 0.2% in 2002. Vanuatu relies on aid, particularly from Australia and New Zealand.

3 The Future

a) Population

One of the problems facing the planet is the continuing rise in population. The population of the planet first reached 2 billion in the 1930s and it took less than 70 years for this population to triple. Predictions estimate that the population will have increased to 7.5 billion by 2020, and that the greatest proportion of this increase will be found in the LEDCs of the world whose fertility rate of 3.1% is double that of the MEDC nations of the world.

Four of the 20 countries with the largest populations identified in 2003 were MEDCs, with the remaining 16 being LEDCs, which, with the exception of Brazil and Mexico, all are located in the continents of Asia and Africa.

Rank	Country	Population (millions)	GNI (US$)
1	China	1289	3,950
2	India	1069	2,820
3	USA	292	34,280
4	Indonesia	220	2,830
5	Brazil	176	7,070
6	Pakistan	149	1,860
7	Bangladesh	147	1,600
8	Russia	146	6,880
9	Nigeria	134	790
10	Japan	128	25,550
11	Mexico	105	8,240
12	Germany	83	25,240
13	Philippines	82	4,070
14	Vietnam	81	2,070
15	Egypt	72	3,560
16	Turkey	71	5,830
17	Ethiopia	71	800
18	Iran	67	5,940
19	Thailand	63	6,230
20	France	60	24,080

Figure 52 The world's largest countries by population in 2003. *Source*: 2003 World Population Data Sheet.

Rank	Country	Population (millions)
1	India	1628
2	China	1394
3	USA	422
4	Pakistan	349
5	Indonesia	316
6	Nigeria	307
7	Bangladesh	255
8	Brazil	221
9	Congo, Democratic Republic of	181
10	Ethiopia	173
11	Mexico	153
12	Philippines	133
13	Egypt	127
14	Russia	119
15	Vietnam	117
16	Japan	101
17	Turkey	98
18	Iran	96
19	Sudan	84
20	Uganda	82

Figure 53 Population predictions for the world's largest countries by population in 2050. *Source*: 2003 World Population Data Sheet.

By 2050, only the MEDC nations of Japan, Russia and the USA will be in the 20 most populous countries and of these only the USA will have seen a population increase. This can be accounted for by a higher fertility rate especially among the large Hispanic population. Like many MEDC nations, Japan and Russia display characteristics of a falling and ageing population, with a greater proportion of their population being outside of the economically productive 16–65 age group.

- In Italy, to offset the fall in population, the government is giving women who have a second child within 18 months of the first child, a tax-free lump sum of money of approximately €1000 per child. The decision to extend this to first-time mothers in 2005 will be at a cost of €500 million.
- Spain is encouraging Argentinian people with Spanish relatives to emigrate and repopulate the rural areas of the country which are experiencing population decline.
- Facing population decline, France has an official policy to increase the birth rate stating that 'the more children born, the better it is for the state's pension system and economy in general'.
- Meanwhile, in the UK, population is increasing by 250,000 people a year with 75% of this total being made up by immigrants. Predictions see a population of 71.3 million by 2050. This has aroused concerns among groups such as the Optimum Population

Trust and conservationists such as David Attenborough, who have urged the government to stabilise the county's population and even allow it to decline naturally to reduce pressures on the environment.

By 2050 it is predicted that the MEDC nations of France and Germany and the LEDC nation of Thailand will be replaced in the 20 most populous nations by the LEDC nations of the Democratic Republic of Congo, The Sudan and Uganda, who are predicted to have experienced a percentage population increase of 220%, 121%, and 226%, respectively, compared to a 7% rise, an 18% fall, and a 15% rise, respectively, for the three countries they will replace.

The United Nations 'National Population Policies' (2001) reports that 76 LEDC countries have policies to reduce their population growth rate. This is in comparison to 20 MEDC countries, mainly in Europe, that have policies to increase their birth rate.

China, India and Pakistan all aim to have stable populations in the foreseeable future:

- China aims to stabilise its population by 2050
- India and Pakistan hope to stem growth by 2045 and 2020, respectively

Sixty per cent of LEDCs have policies aimed at lowering fertility and birth rates in their countries.

b) Established and emerging economies

The UK Office for National Statistics has estimated the value of the country to be £5000 billion. In 2002 there was £61 billion invested in new plant, and only the farming and mining industries recording a fall in employment and investment. The UK has recorded 12 years of economic growth since the spring of 1992. Increasing productivity and economic efficiency also mean that the UK no longer lags behind its main competitors. September 2003 saw the greatest level of manufacturing activity in 16 months and a further upturn in manufacturing is predicted as other MEDC nations emerge from recession.

The US economy saw a 7.2% growth in the third quarter of 2003, although this fell back to 4% in the final quarter of the year. This still left the US economy averaging a 4.5% growth for the year 2003 and predictions indicate that the country has adopted a pattern of economic growth that has had a knock-on effect for other world economies.

Even though the most stable economies are to be found in the MEDCs, economies of certain LEDC nations are developing to the extent that they are becoming attractive locations for companies to relocate to. India, in particular, with its emphasis on a high standard of education, has become an attractive location for outsourcing work

	India	China
Average wage	$1192 per year	$729 per year
Average wage of information technology engineer	$1000 per month	$400 per month
Outsourcing revenue in 2003	$11.36 billion per year	$3.77 billion per year
Estimated outsourcing revenue for 2006	$27.06 billion	$27.54 billion

Figure 54 Comparative wage costs between India and China.

such as customer services. Lower wage costs, the average annual salary being £1200 a year compared to £12,500 in the UK, total operating costs of 30–40% lower than in the UK and fluency in English make outsourcing such industries economic sense. Hence, India in 2003 accounted for 80% of the offshore outsourcing market.

This has led to:

- 28 companies outsourcing 50,000 jobs to India between 2001 and 2003.
- The number of English-language call centres in India rising from 50 to 800 between 2000 and 2003.
- A prediction that 750,000 jobs could go offshore in the next decade, including 50,000 senior management positions.

This structural change in employment patterns is the direct result of technological advances and companies wishing to maximise profit. Relocating aspects of a company to cheaper locations is attractive where face-to-face contact with the general public is unnecessary. The redundancy that this leads to in the original workforce is now a concern for governments and trade unions.

Amicus, the UK's largest private sector trade union, estimates that 200,000 finance sector jobs will be lost in the country by 2008 as their work is transferred to India. Deloitte Consulting estimates that 2 million jobs will be transferred to India by 2008 and the world's 100 major financial institutions will save $138 billion by 2008 by moving their operations offshore.

However, as other emerging nations develop, we may even see work shift from India to locations where even greater savings can be made. China's educated workforce and lower wages make for a viable alternative location.

Countries of eastern Europe that joined the EU in 2004 may also provide cheap alternatives for this work. The Czech Republic and Hungary, in particular, are considered to be suitable locations for such work as the workforce is often fluent not only in English but other Slavic and northern European languages.

4 Impoverished Nations

The wealth of a nation is determined by its natural resources, the education of its people, its infrastructure, its recent history and its ability to trade. If a country is lacking in minerals or agricultural practice stays at a mainly subsistence level, it is unlikely to develop as quickly as a country that is rich in natural resources. This can be seen in both Burkina Faso and Cambodia.

a) Burkina Faso

Burkina Faso in west Africa is one of the poorest countries in the world as a result of its high population density, limited natural resources, including fragile soil, and climatic conditions associated with a desert climate. The average GNI of $1120 hides the extremes that exist – 45% of the population live in poverty. The poorest 10% of the population have access to 2% of the nation's GNI while the top 10% have access to 46.8% according to recent estimates. Adult literacy rates are 6.6% for those aged 15 or over, but this masks the difference between males (36.9%) and females (16.6%).

Approximately 90% of the population are involved in agriculture and it produces 35% of the country's income. Farming is mostly at subsistence levels and the success of the crops depends on whether the limited seasonal rainfall occurs. Burkina Faso's only other natural resource is gold and prospecting has identified further reserves to those already known. This may give the country an additional source of income in the future but as investment in schemes to exploit the reserves would come from overseas, the level of benefit to the country would be relatively low, as present examples prove – the government of Burkina Faso only has a 20% stake in a development sponsored by High River Gold Mines.

Growth of the economy in 2002 was estimated at 4.6%. While this growth is evident in the capital Ouagadougou, the benefits are not being felt by the majority of the population, who are still living in rural poverty. To further grow the economy, a reduction in inflation and the trade deficit, and the encouragement of private enterprise must take place.

b) Cambodia

Cambodia has suffered prolonged periods of war. The invasion in 1978 by Vietnam led to a 20-year period of fighting and civil war that only ceased in 1998 with the surrender of the final Khmer Rouge forces. Such instability led to economic decline and Cambodia's first year of peace in 1999 saw economic growth of 5%. This has been roughly maintained but the very low base from which the economy has started to develop means that the country is still underdeveloped.

The largely rural population live in an environment lacking a basic infrastructure. The agrarian economy is boosted by money generated from tourism and small-scale exploitation of its few mineral resources. The average GNI of $1790 is not evenly distributed and the 80% employed in agriculture receive least with the lowest 10% having access to only 2.9% of the GNI while the top 10% have access to 33.8%.

5 The Needs of an Interdependent World

Over 66% of the world's population lives on less than $1 a day and the consequences of this go beyond the people who this affects directly.

The UN target to assist these peoples and nations is a contribution of 0.7% of GDP of each donor nation. The aid given is used to assist in programmes ranging from:

- providing supplies of clean fresh water
- universal primary education
- health care, including universal access to reproductive health services and spending on AIDS/HIV
- conflict prevention and post-conflict reconstruction
- a fairer international trading system.

In 2000, the largest donor nations were Japan ($13 billion), the USA ($9.6 billion), Germany ($5 billion) and the UK ($4.5 billion). However, these figures do not represent the 0.7% GDP required by the UN and in the case of the UK it only represents 0.31%. This is, however, greater than the average figure for other MEDC nations, which stands at 0.22%. Only five of the smaller European nations, Denmark, Luxembourg, The Netherlands, Norway and Sweden, reach the target figure. A total of $53 billion in aid was given in 2000, a $3 billion decrease on 1999.

The UK is one of the few countries which is actively increasing the amount of overseas aid being given and the government has a good record of targeting worthwhile projects with financial assistance. The country has recognised the interdependency that exists on a global scale, and, by directing aid to the LEDC nations, the aim is to reduce the inequalities that exist. Inequalities that lead to further migration, land degradation and population increase and prevent countries developing a stronger foundation.

There is a need for a stronger commitment from the wealthy nations to their poorer counterparts to reduce the extreme differences currently seen. The World Trade Organisation (WTO) is the only global international organisation dealing with the rules on international trade and its aim is to ensure that trade flows as smoothly as possible. For the LEDC nations, it aims to provide means by which they can increase their trading opportunities and implement improvements in their infrastructure and technical ability. The development of a free-trade zone across the Americas, for example, is seen as a positive move. However, the power and influence of the MEDC nations, especially over trade in agricultural produce (where both the USA and EU heavily subsidise their agriculture), is seen as a barrier to free trade that would benefit the poorer nations.

Giving the LEDC nations greater access to the world market is seen as a way of stimulating growth. The removal of a trade barriers, such as that imposed by the USA on steel imports, is viewed as important as it goes against the principles of 1994 General Agreement on Tariffs and Trade (GATT) agreement limiting such restrictions. It is up to the MEDC nations, particularly the USA, to adhere to the principles of free trade as it is of benefit to the economies of all nations in such an interdependent world.

6 Overall Conclusions

The economic activity of the world is dynamic. Countries vie for dominance and the world ranking among the leading economic nations is constantly changing. The USA dominates the world both politically and economically. Other MEDC and emerging industrialised nations jostle for economic prominence. The UK, in 2004, is in a strong economic position while the economies of its mainland European competitors have been slow to benefit from the economic upturn. The advantage possessed by the UK will require considerable investment by both public and private enterprise to ensure that the advantages possessed by the economy are not lost to competitors.

The cycle of economic activity would indicate that in time other nations, other than those already dominant, will take on new methods of production which will give them the competitive advantage and see them become more of a dominant force in the world economy.

The MEDC nations do need to consider the needs of the least fortunate nations of the world and create a global economy which is seen as fair, competitive and as compassionate as possible so that the world's wealth can better meet the needs of all its people.

Questions

1. What are the direct affects of low levels of income, literacy and health care on a nation's ability to bring about sustainable development?
2. What are the advantages and disadvantages, to both MEDCs and LEDCs, of outsourcing or relocating jobs to LEDCs?
3. How can the MEDC nations assist the LEDC nations to develop? What, in your opinion, are the major issues that need to be addressed?
4. What will be the economic impact of the population changes suggested in Figures 52 and 53?

Bibliography

Cochrane, A., Hamnett, C. and McDowell, L. (Editors), 1981, *City, Economy and Society*, Harper & Row.

Gordon, G., 1986, *Regional Cities in The UK 1890–1980*, Harper & Row.

Guinness, P. and Nagle, G., 2002, *Advanced Geography*, Hodder & Stoughton.

Horsfall, D., 1982, *Manufacturing Industry*, Blackwell.

Knowles, R. and Wareing, J., 1985, *Economic and Social Geography*, Heineman.

Nagle, G., 2000, *Advanced Geography*, Oxford University Press.

Pryce R., 1977, Fundamentals of Human Geography, Unit 2. Open University.

Ross, S., Morgan, J. and Heelas, R., 2000, *Essential AS Geography*, Nelson Thornes.

Smith, D., 1977, *Patterns in Human Geography*, Penguin.

Tidswell, V., 1990, *Pattern and Process in Human Geography*, Unwin Hyman.

Toyne, P., 1984, *D204 Industrial Location, Fundamentals of Human Geography*, Open University.

Websites

A selection of the websites used is listed below.

www.bbc.co.uk/news

www.birminghamuk.com

www.coltoncompany.com/shipbldg/statistics/world.htm

www.forestry.gov.uk

www.dti.gov.uk/energy/coal/uk_industry

www.hie.co.uk

www.hsbc.com

www.legislation.hmso.gov.uk

www.nestle.com

www.odci.gov/cia/publications/factbook/geos/ml.html

www.riotinto.com

www.city.vancouver.bc.ca

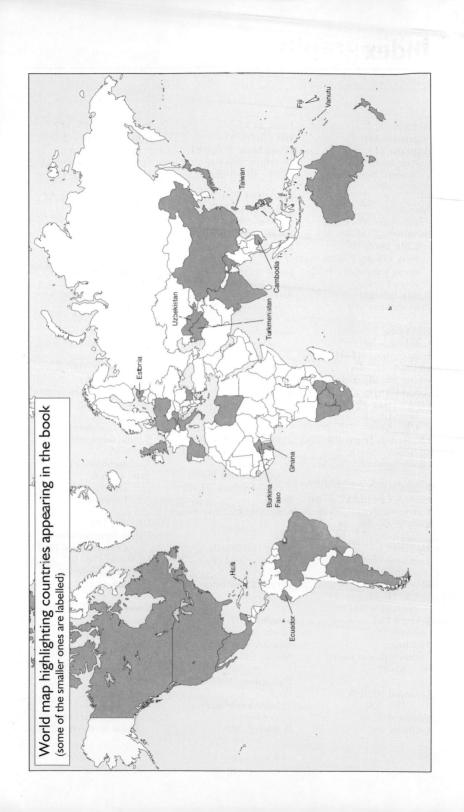

World map highlighting countries appearing in the book
(some of the smaller ones are labelled)

Index